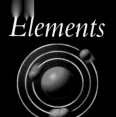

Elements

COPPER
SILVER AND GOLD

Cu

Au

Ag

How to use this book

This book has been carefully developed to help you understand the chemistry of the elements. In it you will find a systematic and comprehensive coverage of the basic qualities of each element. Each two-page entry contains information at various levels of technical content and language, along with definitions of useful technical terms, as shown in the thumbnail diagram to the right. There is a comprehensive glossary of technical terms at the back of the book, along with an extensive index, key facts, an explanation of the Periodic Table, and a description of how to interpret chemical equations.

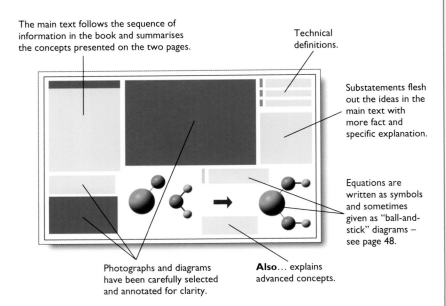

The main text follows the sequence of information in the book and summarises the concepts presented on the two pages.

Technical definitions.

Substatements flesh out the ideas in the main text with more fact and specific explanation.

Equations are written as symbols and sometimes given as "ball-and-stick" diagrams – see page 48.

Photographs and diagrams have been carefully selected and annotated for clarity.

Also... explains advanced concepts.

· ·

An Atlantic Europe Publishing Book

Author
Brian Knapp, BSc, PhD
Project consultant
Keith B. Walshaw, MA, BSc, DPhil
        (Head of Chemistry, Leighton Park School)
Industrial consultant
Jack Brettle, BSc, PhD (Chief Research Scientist, Pilkington plc)
Art Director
Duncan McCrae, BSc
Editor
Elizabeth Walker, BA
Special photography
Ian Gledhill
Illustrations
David Woodroffe
Electronic page make-up
Julie James Graphic Design
Designed and produced by
EARTHSCAPE EDITIONS
Print consultants
Landmark Production Consultants Ltd
Reproduced by
Leo Reprographics
Printed and bound by
Paramount Printing Company Ltd

Suggested cataloguing location
Knapp, Brian
        Copper, silver and gold
        ISBN 1 869860 29 2
        – *Elements* series
540

Acknowledgements
The publishers would like to thank the following for their kind help and advice: *The Copper Development Association*, *Jonathan Frankel* of J. M. Frankel and Associates, *Ian* and *Catherine Gledhill* of Shutters, *Dr Angus W. R. McCrae*, *Rolls-Royce plc* and *Charles Schotman*.

Picture credits
All photographs are from the **Earthscape Editions** photolibrary except the following:
(c=centre t=top b=bottom l=left r=right)
Copper Development Association 11br, 12; **Ian Gledhill** 32bc; courtesy of **Rolls-Royce plc** FRONT COVER, 16/17; by permission of **The Syndics of Cambridge University Library** 37t, 37b, 39b and **ZEFA** 10b, 38c.

Front cover: Copper is used for the stator core of this huge generator used in a power station.
Title page: A spectacular sample of chalcopyrite ore.

First published in 1996 by
Atlantic Europe Publishing Company Limited, Greys Court Farm,
Greys Court, Henley-on-Thames, Oxon, RG9 4PG, UK.

This product is manufactured from sustainable managed forests. For every tree cut down at least one more is planted.

The demonstrations described or illustrated in this book are not for replication. The Publisher cannot accept any responsibility for any accidents or injuries that may result from conducting the experiments described or illustrated in this book.

Contents

Introduction

An element is a substance that cannot be broken down into a simpler substance by any known means. Each of the 92 naturally occurring elements is therefore one of the fundamental materials from which everything in the Universe is made. This book is about copper, silver and gold.

These three elements are often called the "coinage metals" because they are used to make most of the world's coins. One reason for this is that none of the coinage metals is very reactive with other elements, and therefore they are very resistant to corrosion.

Copper

Copper, a soft orangy-coloured metal, was one of the first metals to be used in the ancient world. It has been exploited for at least 7000 years. Its name comes from the Latin, *cuprum*, which means "metal of Cyprus", an island in the Mediterranean Sea where the Romans had large copper mines.

Copper is an excellent conductor of heat and electricity and is found in most of the flexible cables used in the world. Its softness also makes it suitable for tubing for water pipes and central heating systems, because it can be soldered easily and readily bent to fit around corners. Above all, it can be mixed with other metals to make extremely useful alloys such as brass and bronze.

▼ New York's Statue of Liberty has a pleasing green patina, but the copper is otherwise little affected even after a century of exposure to the weather.

Silver

Silver is a white, shiny, heavy metal. Its symbol is Ag, after *argentum*, a Latin word meaning "white and shining". Silver has been sought after since the earliest times and is regarded as a precious metal, just as gems are precious stones. Yet although silver is known for its precious value, only 16% of all the silver used in the world is used for coins and jewellery, while 40% goes to make photographic film. Much of the rest is used in industry and health services. Mirrors for example, are mostly made by silvering the back of glass.

Gold

The chemical symbol for gold is Au, after the Latin word for gold, *aurum*. Gold is one of the rarest elements found on Earth and has been sought out by people since ancient times. One of the reasons for this is that it is a soft metal that nearly always occurs in pure, or native, form. Ancient peoples could thus make use of gold along with silver and copper without special tools and without refining.

Most of the world's gold is in sea water, where it is too dispersed to be collected. On land it is found in veins and in small fragments in river beds and coastal sands. Its discovery has set off many gold rushes throughout the world.

Gold resists corrosion better than almost any other material. It does not tarnish, but remains a bright, lustrous, deep yellow colour indefinitely. In fact this resistance to corrosion has made the metal vital in the electronics industry, where it is used, for example, to make electrical contacts.

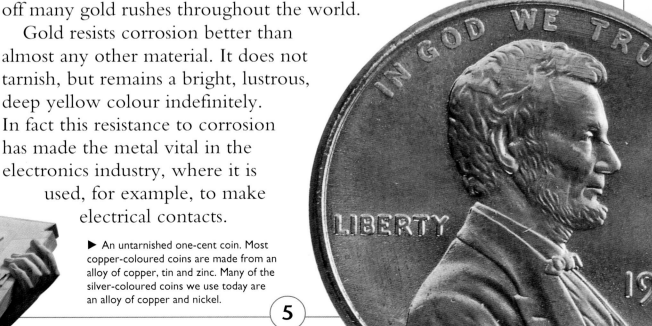

► An untarnished one-cent coin. Most copper-coloured coins are made from an alloy of copper, tin and zinc. Many of the silver-coloured coins we use today are an alloy of copper and nickel.

Copper ores

Copper is a metal that has been deposited from hot sulphur solutions, created as volcanoes were erupting. The hot solutions concentrated the copper by up to a thousand times more than would be normally found in rocks. The resultant enriched rocks are called copper ores.

As the hot fluid made its way from magma chambers through cracks and fissures in the rocks, copper ores were deposited in narrow veins.

The island of Cyprus in the Mediterranean Sea is one such ancient volcanic area. The ancient Romans mined the ore there.

The Romans only mined deposits of native copper, which is copper metal that is not bound up in any compound. This pure form of copper was very easy to work and did not need refining. However, most copper occurs as compounds, especially as sulphides, and because they need to be refined, they have only begun to be used relatively recently.

▲ A piece of chrysocolla.

Chrysocolla

Copper silicate and copper carbonate have a characteristic green colour. Compare this picture to the copper carbonate patina on the Statue of Liberty on page 4. The mineral shown here is called chrysocolla, a copper aluminium silicate.

Chalcopyrite and bornite

Chalcopyrite and bornite are minerals containing both copper and iron sulphides. They are the source of half of the world's copper ores.

Chalcopyrite is a brassy-coloured mineral, while bornite is often a rich peacock-blue (in fact it is often called peacock ore). They are formed during intense volcanic activity, when hot liquids were pushed through fissures in the rocks, cooling and solidifying to give minerals with a high metal content. Silver and gold were formed in much the same way, and for this reason the coinage metals are often mined together.

The sample shown here is made mainly of bornite, but if you look closely you will see the gold speckles of chalcopyrite as well.

▶ A piece of bornite or peacock ore with speckles of chalcopyrite.

magma: the molten rock that forms a balloon-shaped chamber in the rock below a volcano. It is fed by rock moving upwards from below the crust.

native metal: a pure form of a metal, not combined as a compound. Native metal is more common in poorly reactive elements than in those that are very reactive.

ore: a rock containing enough of a useful substance to make mining it worthwhile.

refining: separating a mixture into the simpler substances of which it is made. In the case of a rock, it means the extraction of the metal that is mixed up in the rock.

silicate: a compound containing silicon and oxygen (known as silica).

sulphide: a sulphur compound that contains no oxygen.

vein: a mineral deposit different from, and usually cutting across, the surrounding rocks. Most mineral and metal-bearing veins are deposits filling fractures.

▶ A piece of native copper.

Native copper

Copper is not a very reactive element. Thus, like silver and gold, which are also slow to react chemically, it is sometimes found in pure form. A natural occurrence of pure copper is called native copper. The shape reflects the deep underground fissures in which it was originally deposited and is known as a dendritic pattern.

The largest piece of native copper ever found was in Minesota Mine, Michigan, USA. It weighed over 500 tonnes.

◀ This is banded malachite, a copper carbonate and a useful ore.

7

Reducing copper oxide

Copper is mainly found in the form of a black ore, copper oxide, or a brassy-coloured ore, copper sulphide. In both cases the metal has to be separated from its compound.

The demonstration on this page shows how copper can be extracted from its compound. The ore is copper oxide, a compound of copper and oxygen. To obtain pure copper the oxygen has to be removed, using a process called reduction.

The reducing agent used here is carbon monoxide gas, which is colourless but inflammable. The reaction produces carbon dioxide gas, which is also colourless, but does not burn. So the key to watching this sequence is to look for where the flame appears and disappears. In this way you can tell which gas is in the tube!

❶▼ The black copper oxide is placed in a special glass tube with a small hole near the rounded end. At the start, the tube is full of air. This is swept away by pumping in carbon monoxide gas.

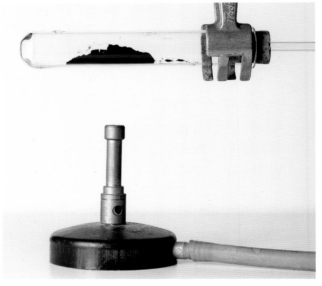

The black copper oxide is heated with a Bunsen burner.

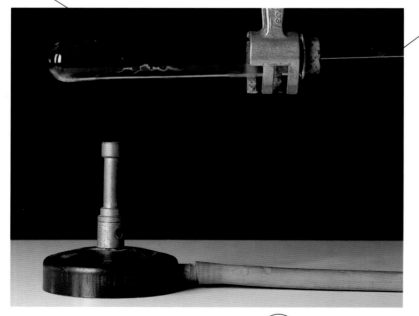

Carbon monoxide gas is passed in through this tube.

❷◄ The copper oxide continues to be reduced, and the oxygen combines with the carbon monoxide to form carbon dioxide. Carbon dioxide does not ignite, so the flame goes out.

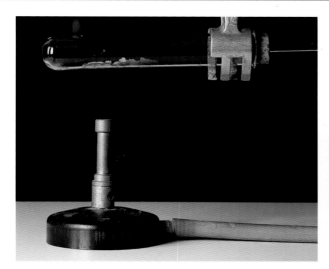

oxide: a compound that includes oxygen and one other element.

reduction: the removal of oxygen from a substance.

Close-up of the end of the tube showing the carbon monoxide gas burning with a blue flame.

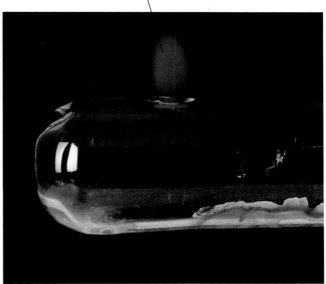

❸▲▶ Reactions between elements and compounds are often very slow at room temperature. To speed up the rate of reaction, the copper oxide is heated using the flame from a Bunsen burner.

The blue flame coming from the small hole in the tube is produced by burning carbon monoxide gas. Notice that the copper oxide is glowing orange on the surface, which shows that the oxygen has been removed.

❹▲ Eventually all of the oxygen is removed from the copper oxide powder, and only copper is left.

EQUATION: Reduction of copper oxide to copper

Copper oxide + carbon monoxide ⇨ copper + carbon dioxide

$$CuO(s) \quad + \quad CO(g) \quad ⇨ \quad Cu(s) \quad + \quad CO_2(g)$$

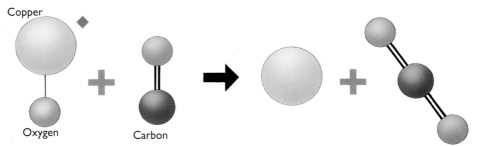

Copper

Oxygen

Carbon

Mining and smelting copper ores

About nine-tenths of the world's reserves of copper are found in just four areas: the great Basin of the western United States, central Canada, the Andes regions of Peru and Chile, and Zambia. In each case the extraction of copper is of crucial importance to the country. In the case of Zambia the reserves are the mainstay of the country's economy, with a chain of major cities making up the "Copper Belt". The largest deposit of copper in the world is at Chuquicamata, Chile, but the largest refiner of copper is the United States, which also boasts the world's largest copper mine, in Utah.

The amount of copper in the ground is relatively small and most of it occurs in low-grade ores that have to be processed twice to extract the copper. This is why it is important to reuse as much copper as possible, and why about one-third of copper consumed in most industrial countries is recycled from scrap.

Mining

Over 90% of the world's copper ore is obtained by strip mining in vast open-cast pits. Large blast holes are drilled in the ore, the material is blasted loose, and is then put into dump-trucks and taken to the enriching plant.

▶ Copper ores can be mined with as little as about one-half of 1% copper content. This picture, from Arizona, USA, shows one of the world's most important copper mines.

The total world reserve of copper is just over 200 million metric tons. It will provide enough copper to last about 50 years.

Concentration

Most ore contains only about 1% metal, so the ore must be concentrated before it is sent for smelting and refining. This is done by first pulverising the ore, then separating it by flotation, and finally drying it. At the end of this process the metal to ore ratio is about one to three.

The ore flotation method

The flotation method is a way of separating ore mineral from gangue (rock that contains no metal), thus enriching the ore before final processing.

gangue: the unwanted material in an ore.

The ores to be enriched are first ground to a fine grit, which still contains particles of copper mixed up with gangue. To separate the copper particles from the gangue, the grit is introduced to a bath of water containing a foaming agent, which produces a kind of bubble bath combined with a special oil-based chemical that makes the copper particles water-repellent.

When jets of air are forced up through the bath, the water-repellent copper particles are picked up by the bubbles of foam and float to the surface, making a froth. The froth is skimmed off the surface, and the enriched ore is taken away for refining.

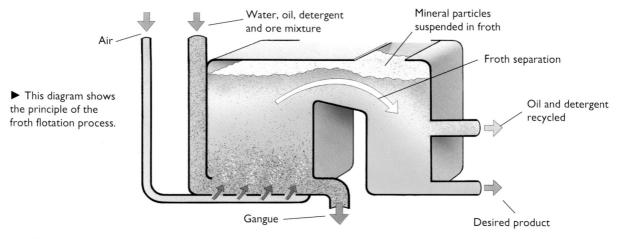

Air

Water, oil, detergent and ore mixture

Mineral particles suspended in froth

Froth separation

Oil and detergent recycled

► This diagram shows the principle of the froth flotation process.

Gangue

Desired product

Smelting

Most copper ores are difficult to refine, although smelting removes most of the impurities from the enriched ore. The easiest ores to smelt are the copper oxides. Carbon monoxide gas needed to reduce the ore is produced by heating coke, a source of carbon, and feeding in a jet of air. This carbon and oxygen from the air react to form carbon monoxide, which reduces the copper oxide to copper. The copper can then be tapped from the base of the furnace.

Copper sulphides present greater difficulties. The copper is removed by heating and the process may require the introduction of oxygen. At the same time the sulphur is oxidised to sulphur dioxide gas. Since sulphur dioxide is one of the major contributors to acid rain, many modern factories recover as much of the gas as possible to be used in the production of sulphuric acid.

EQUATION: Reduction of copper ore containing both oxides and sulphides by heating

Copper sulphide + copper oxide ⇨ copper + sulphur dioxide

$$CuS(s) \quad + \quad 2CuO(s) \quad ⇨ \quad 3Cu(s) \quad + \quad SO_2(g)$$

▲ Molten copper being cast into ingots for later use. A continuously rotating wheel is used for this process.

Electrical refining of copper

Even the best of chemical reactions cannot completely remove all of the impurities in a metal, so ores refined in a furnace do not produce pure metals. This is why many metals are refined to their final stage of purity by electrical means in a process called electrolysis.

Impure copper from the furnace is used as one of the electrodes of an electrolysis cell. The other electrode is made from a thin sheet of pure copper. The copper is then refined by placing the two electrodes in a copper sulphate bath and passing a current between them. The impure copper on the anode corrodes, and copper ions pass through the electrolyte, collecting on the cathode sheet as pure copper.

When the cathode has acquired a sufficient thickness of pure copper, it is lifted from the electrolysis cell and replaced with a new electrode. Similarly, when the anode has corroded completely away, it is replaced with a new ingot of smelted metal. The cathodes are then melted down and made into wire and sheet metal.

The laboratory demonstration of electrolysis and the giant industrial equivalent are shown on these pages.

► Copper cathodes, heavily plated with copper, are lifted from the electrolytic cells.

Laboratory electrolysis

A demonstration of electrolysis can be done using a beaker and two copper strips. A dry battery serves as the source of electrical current. The electrolyte is reagent quality copper sulphate solution.

Signs of corrosion and plating are evident within minutes, and a completely corroded strip can be produced within a day or so.

electrolysis: an electrical–chemical process that uses an electric current to cause the break up of a compound and the movement of metal ions in a solution. The process happens in many natural situations (as for example in rusting) and is also commonly used in industry for purifying (refining) metals or for plating metal objects with a fine, even metal coating.

electrolyte: a solution that conducts electricity.

slag: a mixture of substances that are waste products of a furnace.

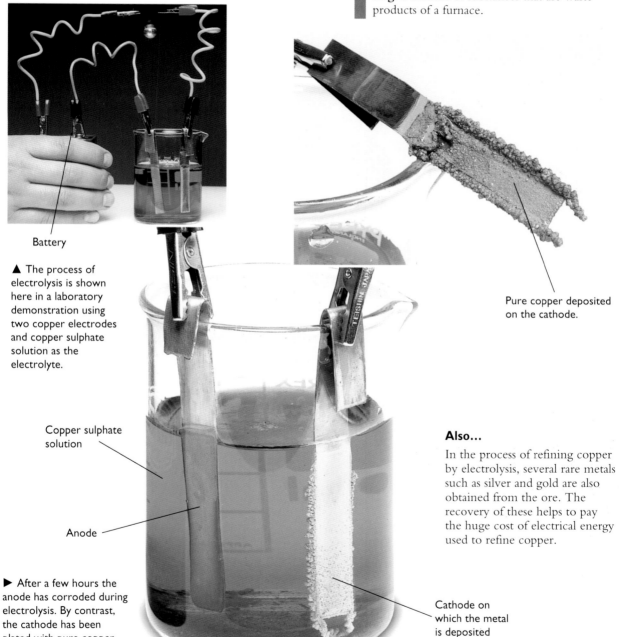

Battery

▲ The process of electrolysis is shown here in a laboratory demonstration using two copper electrodes and copper sulphate solution as the electrolyte.

Pure copper deposited on the cathode.

Copper sulphate solution

Anode

Also...

In the process of refining copper by electrolysis, several rare metals such as silver and gold are also obtained from the ore. The recovery of these helps to pay the huge cost of electrical energy used to refine copper.

► After a few hours the anode has corroded during electrolysis. By contrast, the cathode has been plated with pure copper.

Cathode on which the metal is deposited

Copper as a metal

About nine million tonnes of copper are used every year in a wide variety of ways. About half of all copper is used in the electrical industry (see next page). Copper is also used for water pipes, roofing, locks and hinges, coins, and vehicle radiators.

Many of these uses have come about because, of all the common metals, copper is outstanding in its resistance to attack by oxygen and water. Copper only changes to copper oxide when the temperature reaches 300°C. It is not corroded by water or steam, which is why it can be used for hot and cold water systems and central heating and air conditioning systems. It is not even affected by most dilute acids, although concentrated nitric acid reacts violently with copper, as shown below.

▲ Heated copper compounds produce a characteristic green–blue flame.

◄◄ The picture on the far left shows fuming concentrated nitric acid being poured on to copper turnings (finely divided copper that provides a large surface area for fast reaction). The picture on the near left shows the resulting reaction. The copper is transformed into blue copper nitrate and large amounts of nitrogen dioxide are given off.

EQUATION: Copper and fuming nitric acid

Copper turnings + fuming nitric acid ⇨ copper nitrate + water + nitrogen dioxide

$$Cu(s) + 4HNO_3(l) \Rightarrow Cu(NO_3)_2(s) + 2H_2O(l) + 2NO_2(g)$$
blue

The reactivity of copper

Metals can be arranged in a list, with the most reactive at the top and the least reactive at the bottom. Many metals are subject to corrosion when placed in damp air or damp soil. The most vulnerable of all are the most reactive elements.

Copper comes near the bottom of the reactivity series because it is only slightly reactive. This low reactivity means that copper objects can be placed out in exposed locations without fear that they will corrode away.

corrosion: the *slow* decay of a substance resulting from contact with gases and liquids in the environment. The term is often applied to metals.

REACTIVITY SERIES	
Element	Reactivity
potassium	*most reactive*
sodium	
calcium	
magnesium	
aluminium	
manganese	
chromium	
zinc	
iron	
cadmium	
tin	
lead	
copper	
mercury	
silver	
gold	
platinum	*least reactive*

► Copper is a good conductor of heat. This makes for efficient use of energy and precise control of the cooking temperature when copper is used for cooking implements such as this jam-making pan.

◄ Copper is widely used for decorative metalware, either as pure copper or as an alloy such as brass.

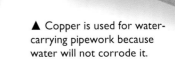

▲ Copper is used for water-carrying pipework because water will not corrode it.

Copper as a conductor

All metals have the ability to transfer, or conduct, heat and electricity; however, copper is among the most efficient at conducting both.

An electric current is simply a flow of electrons. Metals can conduct electricity because they are made up of a "honeycomb" (known as a lattice) of positively charged ions in a "sea" of electrons. In the case of copper, these electrons are not bound to any one copper ion but can move freely. This is what makes copper such a good conductor of electricity.

When heat is applied to copper, the atoms of the metal vibrate and pass energy across the honeycomb. Copper conducts heat better than most metals because of the arrangement of its atoms.

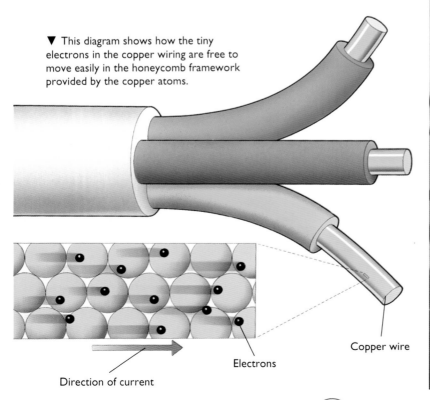

▼ This diagram shows how the tiny electrons in the copper wiring are free to move easily in the honeycomb framework provided by the copper atoms.

Copper wire

Electrons

Direction of current

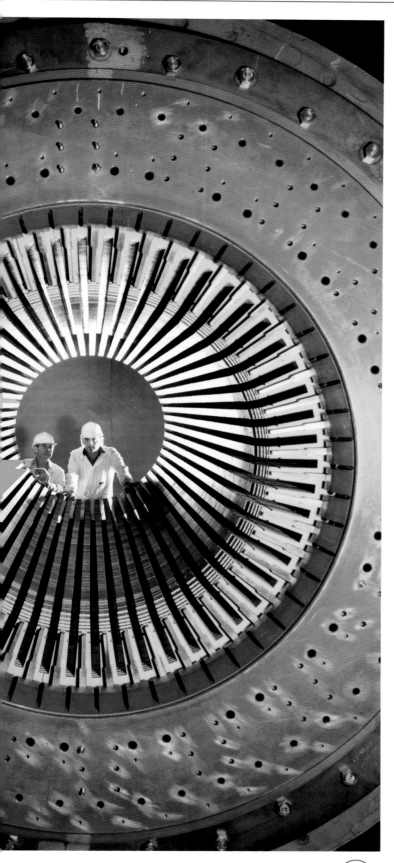

electron: a tiny, negatively charged particle that is part of an atom. The flow of electrons through a solid material such as a wire produces an electric current.

ion: an atom, or group of atoms, that has gained or lost one or more electrons and so developed an electrical charge.

◀ Looking through the inside of a large electrical motor. The copper wire carries the electrical current that generates the magnetic field that, in turn, causes the drive shaft of the motor to turn.

Copper in printed circuits

A printed circuit is made from a rigid baseboard of insulating material with conductors stuck to its surfaces. To make a printed circuit, a sheet of copper foil is stuck to both sides of the baseboard. The copper foil is then sprayed with a film of light-sensitive (silver-based) material. A mask is made up of the connections that are required and placed on the film. The film is then exposed to light and the board put into a photographic developer, which removes all undeveloped parts of the film.

Next the board is placed in an acid solution that dissolves away all parts of the copper not protected by the developed film. This leaves behind the pattern of connections on the board, ready for the electrical components to be added.

▼ A printed circuit board showing the conducting circuit of fine copper connecting the wide variety of electronic devices.

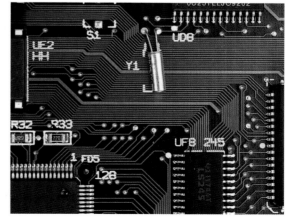

Copper alloys: brass

An alloy is a mixture of metals. Copper forms alloys more easily than most other metals. Each of the alloying metals gives the alloy its own special properties. Some metals make the alloy stronger, others change its colour, make it easier to machine or make it even more resistant to corrosion or wear. The metals most often alloyed with copper are shown on the next few pages.

Brass

Brass is one of the most widely used alloys. It is mainly copper, alloyed with between 5 and 40% zinc. Brass is often used for corrosion-resistant decorative purposes such as door furniture. It is much harder and stronger than copper and it will machine well.

The most common mixture of brass contains 36% zinc and is known as common brass. The properties of brass can be altered significantly by adding small quantities of other elements. Those most commonly used are lead, tin, aluminium, manganese, iron, nickel, arsenic and silicon. For example, by adding up to 3% lead the machinability of brass can be improved significantly.

Copper-rich brasses have special uses, such as making the percussion caps of ammunition; those with between 10 and 20% zinc are called gilding metals and are used for decorative brasswork and jewellery. This form of brass will take an enamel well and is easy to braze.

As the amount of zinc is increased still further, the brass develops the property of being easily shaped when hot. This material is used to make inexpensive, but complex engineering shapes that are easy to machine.

However, even higher proportions of zinc make the alloy more susceptible to corrosion when the brass is placed in water. To counteract this problem, arsenic is added to the alloy.

Tin can also be added to brass to improve its corrosion resistance, and tin–zinc–copper brasses, in which there is 1% tin, are known as admiralty brass because of their suitability for use on ships.

▼▶ Brass is usually made from about 64% copper and 36% zinc. Adjusting the proportions produces very different properties; for example, the alloy becomes harder as more zinc is added. Other metals can be added to give additional qualities, as the diagram on the right illustrates.

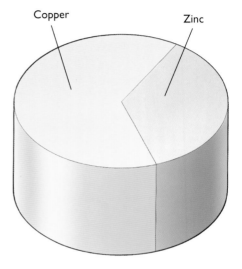

Copper   Zinc

▼ A jug made of brass and copper.

Brass top and handle. The tarnished surface has been cleaned using a solvent.

This part of the jug is made of copper.

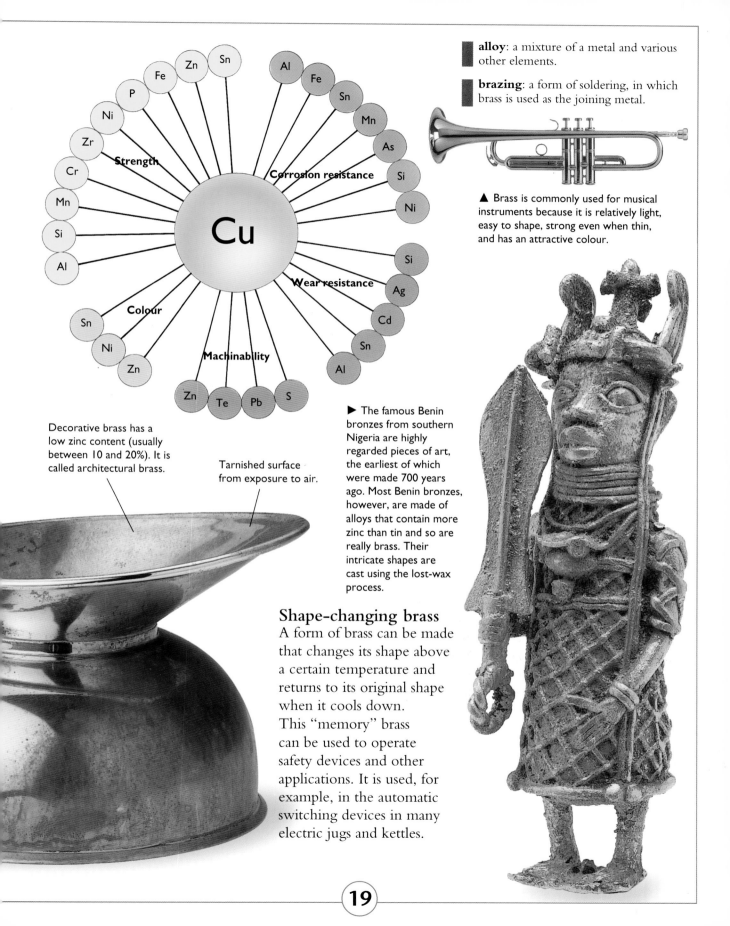

Strength

Fe, Zn, Sn, P, Ni, Zr, Cr, Mn, Si, Al

Corrosion resistance

Al, Fe, Sn, Mn, As, Si, Ni

Cu

Wear resistance

Si, Ag, Cd, Sn, Al

Colour

Sn, Ni, Zn

Machinability

Zn, Te, Pb, S

alloy: a mixture of a metal and various other elements.

brazing: a form of soldering, in which brass is used as the joining metal.

▲ Brass is commonly used for musical instruments because it is relatively light, easy to shape, strong even when thin, and has an attractive colour.

Decorative brass has a low zinc content (usually between 10 and 20%). It is called architectural brass.

Tarnished surface from exposure to air.

► The famous Benin bronzes from southern Nigeria are highly regarded pieces of art, the earliest of which were made 700 years ago. Most Benin bronzes, however, are made of alloys that contain more zinc than tin and so are really brass. Their intricate shapes are cast using the lost-wax process.

Shape-changing brass

A form of brass can be made that changes its shape above a certain temperature and returns to its original shape when it cools down. This "memory" brass can be used to operate safety devices and other applications. It is used, for example, in the automatic switching devices in many electric jugs and kettles.

Copper alloys: bronze

Bronze is an alloy of copper that is significantly different from brass. Bronze is a copper alloy with tin as its major secondary constituent (brass is an alloy with zinc as described on page 18).

Bronze has been used since ancient times for decorative metal objects and also for coins. It was one of the earliest metal alloys used, giving rise to the first metal-working age, known as the Bronze Age, over 3000 years ago. Bronze Age people, however, did not know about alloying (mixing) metals, but used copper ores that naturally contained tin impurities.

▲▼ Bronze is usually made from about 78% copper and 12% tin. The less tin, the softer the metal will be. Usually, no more than 25% tin is added.

Adding zinc and lead to the bronze alloy produces a material that is much more suitable for casting.

An alloy with about nine-tenths copper and equal proportions of the other metals is called gunmetal. It was commonly used in cannons, not only for its corrosion resistance but also for its machinability. Above is a cannon that was used at Gettysburg during the American Civil War.

Wrought bronze

Wrought bronze, also known as phosphor bronze, has a low tin content (perhaps 8% or less) and 0.3% phosphorus. It is commonly used for bearings in machines or engines where shafts continually rotate. It is also used for coins.

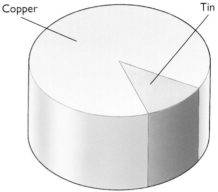

Copper — Tin

◀ A bronze coin with a portrait of the Roman emperor Hadrian, who reigned between 117 and 138 A.D. This particular coin was minted sometime between 126 and 138 A.D. Despite being in damp soil for more than 1850 years the coin still shows a lot of detail from the original casting.

Gastight or porous bronzes

If lead is used in bronze alloys, the alloy will run into any pores created as the molten alloy sets. Leaded bronze is thus gastight and can be used for pressure vessels.

On the other hand, bronze can be made intentionally porous, so that it can be used as a filter. For this purpose tin and copper powders are mixed, then heated in the absence of air. The result is a porous bronze.

Bell-making bronze

There is a wide range of specialised bronzes, each one having its own distinctive properties. Up to 12% tin is normally used in making a bronze, but up to 20% is used for bell-making bronze, which is little used for engineering. The brittleness of bell-making bronze makes the bells liable to crack. However, this disadvantage is outweighed by the particularly sonorous tones made by the metal.

▼ The famous Liberty Bell is kept in Philadelphia, Pennsylvania, USA. Bell metals like that used for the Liberty Bell contain about 20–25% tin and are very hard, giving them the appropriate resonance. However, this also makes them very brittle. The crack in the Liberty Bell is so bad that it cannot be rung.

porous: a material containing many small holes or cracks. Quite often the pores are connected, and liquids, such as water or oil, can move through them.

▼ An English bronze bell foundry at work in the 18th century. The furnace can be seen in the background. The molten bronze was poured into a mould and allowed to solidify before being taken out and filed, to shape and tune the bell.

Corrosion resistance

Silicon bronze is used in places where there is a great danger of corrosion, such as in a chemical works. This alloy contains up to 3% silicon. Bronzes made of alloys of aluminium and copper have similar properties.

▲ The Chinese, Romans and Greeks used bronze extensively for casting sculptures, and it is still used for this purpose today, as shown by the Bucking Bronco statue from Denver, Colorado, USA, above. However, bronze is now considered to be too expensive and labour intensive to use in comparison with other alloys and metals available.

Copper in the environment

Both plants and animals need a certain amount of copper as part of their diets. In animals, copper is used in the body to help release energy and is usually found in protein. It is required as a micronutrient, that is, it is vital but is only needed in very small amounts. Too little or too much can be dangerous. Copper is a natural part of most soils and is used by plants too as a micronutrient to aid growth.

▲ Where the copper content of a soil is low, plants can suffer from poor growth and become much more prone to disease. The different rates of growth from different amounts of copper can be seen in this picture. The wheat on the left has an adequate amount of copper; the one on the right is copper-deficient.

▼ Copper-based fungicide is commonly sprayed on fruit trees.

Copper in gardens and farms

The main copper–based fertiliser is copper sulphate, which can be applied to the soil or to the leaves of growing plants as a spray.

Plants are able to benefit from a spray of copper, but many pests are not. Both copper nitrate and copper sulphate poison fungi (moulds), algae and bacteria. For this reason copper salts are used extensively by farmers and gardeners.

Copper sulphate has traditionally been used to make a fungicide called Bordeaux mixture, once used on the vineyards of the Bordeaux region of France. Seeds are also dipped in copper sulphate solution to prevent disease.

micronutrient: an element that the body requires in small amounts. Another term is trace element.

protein: molecules that help to build tissue and bone and therefore make new body cells. Proteins contain amino acids.

Copper and pollution

Copper tailings, the waste products of copper mines, contain high concentrations of copper. At these levels they are poisonous to plants, and spoil tips remain bare of vegetation. Care has to be taken to make sure contaminated water does not reach nearby rivers.

◄ This building is being sprayed with a copper-based fungicide during construction to help protect it against attack by mould.

Copper compounds as preservatives

In the home, and on exposed timber such as fences, copper salts can be used to prevent mould growing in damp areas. Copper sulphate is routinely added to wallpaper paste for this purpose.

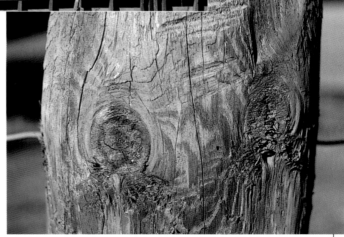

▲ Fence posts and other wood intended for prolonged outdoor use is pressure-treated with copper and other compounds to prevent rotting.

Copper sulphate

One of the best known compounds of copper is copper sulphate (formerly known as blue vitriol because of its deep blue colour). It is prepared by reacting copper oxide or copper carbonate with warm, dilute sulphuric acid. The solution then turns a translucent blue.

To make copper sulphate crystals, the solution is evaporated until it is saturated (it can hold no more copper sulphate in solution). Large crystals of copper sulphate can be grown in this solution by suspending a small "seed" crystal in the saturated solution, around which larger crystals will form. Smaller crystals are produced by evaporating the solution to a small volume. Each technique is shown in the pictures on this page.

❶◄ Dilute sulphuric acid is poured on to (black) copper oxide

❷► The reaction occurs as the reagents are stirred.

EQUATION: Producing copper sulphate

Copper oxide + dilute sulphuric acid ⇨ copper sulphate + water

$$CuO(s) \quad + \quad H_2SO_4(aq) \quad ⇨ \quad CuSO_4(aq) \quad + \quad H_2O(l)$$

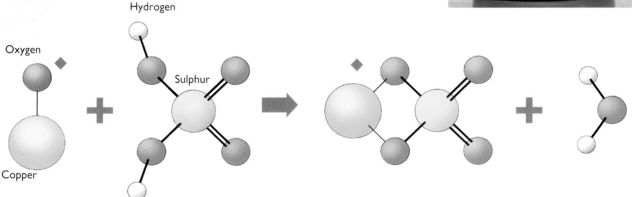

Hydrogen

Oxygen

Sulphur

Copper

3 ◀ The solution soon clears and becomes bright blue. This is copper sulphate dissolved in water.

anhydrous: a substance from which water has been removed by heating. Many hydrated salts are crystalline. When they are heated and the water is driven off, the material changes to an anhydrous powder.

hydrate: a solid compound in crystalline form that contains molecular water. Hydrates commonly form when a solution of a soluble salt is evaporated. The water that forms part of a hydrate crystal is known as the "water of crystallization". It can usually be removed by heating, leaving an anhydrous salt.

reagent: a starting material for a reaction.

◀▼ Copper sulphate crystals like these of varying size can be grown from a saturated copper sulphate solution.

▲▶ Crystals of hydrated copper sulphate (above) are deep blue. However, when the water is driven off, the substance changes to a very light blue powder, called anhydrous copper sulphate (right).

Copper colours

A characteristic of many metals is that their compounds produce a variety of coloured substances, usually quite unlike the colour of the metals in their native state.

Copper forms several such compounds. For example, copper oxide is black; copper carbonate is green; copper sulphate and copper nitrate are blue. Copper sulphide may be black or, if it is an ore, brassy yellow, depending on the amount of iron it contains.

Some examples of these colours are shown on this page, but other examples can be found on page 6.

❶▲ Nitric acid is being poured onto black copper oxide.

EQUATION: Producing copper nitrate

Copper oxide + nitric acid ⇨ copper nitrate + water

$$CuO(s) + 2HNO_3(l) \Rightarrow Cu(NO_3)_2(s) + H_2O(l)$$

▶ This represents copper nitrate.

Nitrogen

Oxygen

Copper

❷▼ Crystals of copper nitrate form. They are a brighter blue than copper sulphate.

▼ Pure copper is an orangy colour, as shown by these coppered nails.

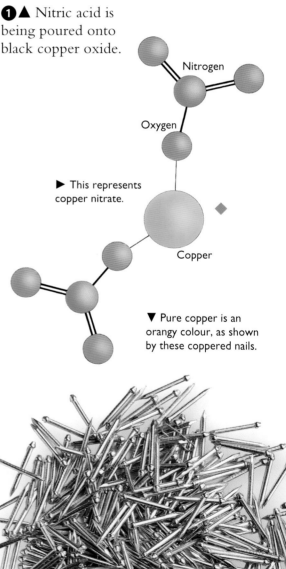

Green copper coating

Copper corrodes in moist air due to the combined effects of copper, water, oxygen and carbon dioxide from the air.

This makes a pleasing green copper carbonate coating known as a patina, or incrustation. A patina protects the surface of a substance from further attack, and forms on copper, brass and bronze objects.

gelatinous: a term meaning made with water. Because a gelatinous precipitate is mostly water, it is of a similar density to water and will float or lie suspended in the liquid.

patina: a surface coating that develops on metals and protects them from further corrosion.

EQUATION: How copper develops a green colour when exposed to the environment

Copper + carbon dioxide + oxygen (dissolved in water) ⇨ copper carbonate

$$2Cu(s) \quad + \quad 2CO_2(g) \quad + \quad O_2(aq) \quad ⇨ \quad 2CuCO_3(s)$$

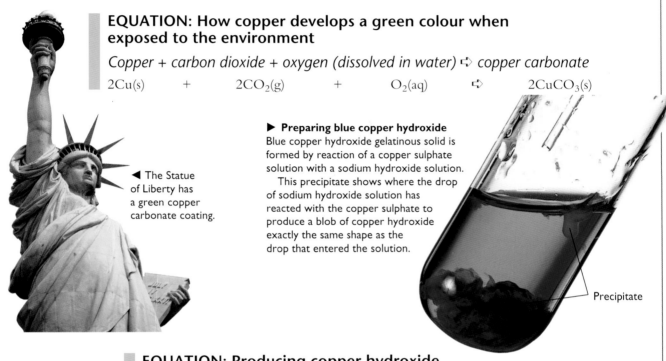

◀ The Statue of Liberty has a green copper carbonate coating.

▶ **Preparing blue copper hydroxide**
Blue copper hydroxide gelatinous solid is formed by reaction of a copper sulphate solution with a sodium hydroxide solution.

This precipitate shows where the drop of sodium hydroxide solution has reacted with the copper sulphate to produce a blob of copper hydroxide exactly the same shape as the drop that entered the solution.

Precipitate

EQUATION: Producing copper hydroxide

Copper sulphate + sodium hydroxide ⇨ copper hydroxide + sodium sulphate

$$CuSO_4(aq) \quad + \quad 2NaOH(aq) \quad ⇨ \quad Cu(OH)_2(s) \quad + \quad Na_2SO_4(aq)$$
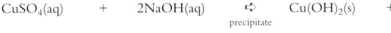
precipitate

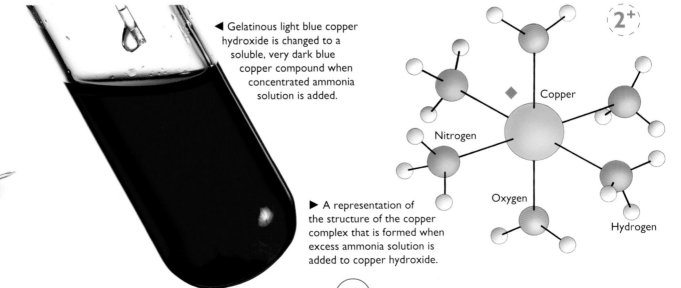

◀ Gelatinous light blue copper hydroxide is changed to a soluble, very dark blue copper compound when concentrated ammonia solution is added.

2^+

Copper

Nitrogen

Oxygen

Hydrogen

▶ A representation of the structure of the copper complex that is formed when excess ammonia solution is added to copper hydroxide.

Silver

Silver lies between copper and gold as one of the softest metals. It is the best conductor of heat and electricity of all the metals, but its softness and relative rarity mean that it has not been put into widespread use.

Silver, gold, platinum and mercury make up the noble metals. They share the common property that they do not oxidise readily when heated, and they will not dissolve in most mineral acids. Silver has been treasured since ancient times, and it is called a precious metal, as are gold, platinum, iridium and palladium.

The main use of silver in the past has been for coins and jewellery. Even here softness is a problem, and silver jewellery and coins are actually alloys of silver and copper. In fact, silver coins are at least one-tenth copper.

▼ Silver coins have been in existence for thousands of years. This is an example of a Greek tetradrachma showing Apollo. It was minted between 261 and 246 B.C.

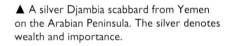

▲ A silver Djambia scabbard from Yemen on the Arabian Peninsula. The silver denotes wealth and importance.

Sterling silver
Silver jewellery, cutlery and serving dishes are made of sterling silver. It consists of 92.5% silver and 7.5% copper, the copper being used to make the silver harder and more able to withstand the occasional knock without denting.

The reactivity of silver

Metals can be arranged in a list called the reactivity series, with the most reactive at the top and the least reactive at the bottom.

Silver is near the bottom of the reactivity series because it is only very slightly reactive. This low reactivity means that silver compounds are rarely produced in nature, and silver has very limited uses in chemistry. It will not react with the air to form oxides, although it does react with polluted air to form silver sulphide. This is, in fact, the chemical composition of the tarnishing seen on silver jewellery, cutlery, and decorative ware, as described below.

alloy: a mixture of a metal and various other elements.

noble metal: silver, gold, platinum, and mercury. These are the least reactive metals.

precious metal: silver, gold, platinum, iridium, and palladium. Each is prized for its rarity. This category is the equivalent of precious stones, or gemstones, for minerals.

reactivity: the tendency of a substance to react with other substances. The term is most widely used in comparing the reactivity of metals. Metals are arranged in a reactivity series.

REACTIVITY SERIES

Element	Reactivity
potassium sodium calcium magnesium aluminium manganese chromium zinc iron cadmium tin lead copper mercury **silver** gold platinum	most reactive least reactive

▼ Silver metal reacts with hydrogen sulphide gas in the air to produce black silver sulphide.

Tarnish

Tarnish is the dark brown or black film that develops slowly on silverware, especially in industrial cities. Silver does not react with oxygen, so tarnish is not an oxide coating. Rather, it is a reaction of the silver with hydrogen sulphide in the air, especially the air near industrial cities. The result is a black film of silver sulphide as shown on the fork above and to the right.

Some silver tableware can tarnish (for example, the ends of forks) because some foods contain hydrogen sulphide. Hard boiled eggs are an example. The hydrogen sulphide effect can also be seen in the dark ring around the yolk.

EQUATION: Tarnishing of silver

Silver + hydrogen sulphide + oxygen ⇨ silver sulphide + water

$$4Ag(s) + 2H_2S(g) + O_2(g) \Rightarrow 2Ag_2S(s) + 2H_2O(l)$$

Silver and silver ores

Silver is a rare element, being 68th in the elements of the earth's crust. Because it reacts poorly, it is sometimes found as native (pure) silver metal. This allowed the Egyptians, for example, to use silver nearly 5000 years ago. Similar native deposits, called lodes, have been found in the Americas. This attracted Spanish colonists in the 17th and 18th centuries, and the discovery of the Comstock Lode in Nevada in the 19th century caused a silver rush.

Outside these rare native deposits, silver is more commonly found as silver sulphide. Much silver is also found associated with other metals such as zinc and copper, and is recovered as a byproduct of refining these more plentiful metals.

Hydrothermal veins and native deposits

Native metals are found only in places that were once so hot that the metals existed in molten form. It is quite common to find deposits of all of the coinage metals together. This is because they are all products of hot liquids created during intense volcanic activity.

Deep below volcanoes lie their source chambers, full of liquid rock (called magma) that has forced its way up from deep within the Earth's crust. At the end of the volcanic activity the magma chambers begin to cool and many of the constituents crystallise out. At the same time, hot acidic liquids are produced that flow out from the magma into fissures in the surrounding rock. Here the hot solutions cool and the various dissolved compounds solidify in fissures called veins.

Veins contain a variety of minerals, including metal compounds and native metals. These are the rich "lodes" that prospectors have sought through the centuries.

Less concentrated deposits have been produced by sedimentary processes, and the ores, though much more extensive, are far less rich in metal.

◄ This diagram shows how hydrothermal veins are related to the magma source, which subsequently cools to granite rock. Erosion often strips the surface rocks away, leaving deposits in a ring pattern around the granite.

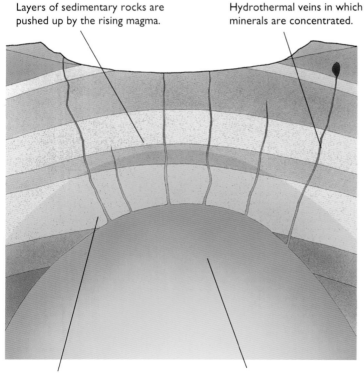

Layers of sedimentary rocks are pushed up by the rising magma.

Hydrothermal veins in which minerals are concentrated.

Rocks around the hot magma chamber are metamorphosed, or changed.

Magma from below the Earth's crust initially heats the surrounding rocks but eventually cools to form granite.

Extraction and refining

Silver is present in small concentrations along with several other metals, such as copper, lead and zinc and it is usually extracted while these metals are being refined.

During electrical refining of copper, for example, the silver settles out at the bottom of the tank in which copper is being refined by electrolysis. The silver is dissolved out of the residue using concentrated nitric acid. The solution is then reacted with iron sulphate, and the silver comes out of solution as a precipitate. It is then further refined by electrolysis.

Silver also dissolves in solutions of sodium or potassium cyanide, and these substances can be used for chemical extraction. The reaction produces a solution containing silver cyanide to which zinc is added causing the silver metal to precipitate out. This method was developed just in time to be used to extract silver from the Comstock Lode.

Because cyanide is so dangerous, the cyanide process is little used today. Instead, a flotation process is used (see page 10). Jets of water create bubbles in tanks of ore powder containing a frothing agent. The silver is taken way in the froth and is then refined by electrical means.

Because so much silver is used in film-making, attempts are made to recover the silver on the film and recycle it during film processing. About one fifth of the silver used in film-making is recovered. This is done by burning the film and then refining the silver by electrical means.

gangue: the unwanted material in an ore.

hydrothermal: a process in which hot water is involved. It is usually used in the context of rock formation because hot water and other fluids sent outwards from liquid magmas are important carriers of metals and the minerals that form gemstones.

lode: a deposit in which a number of veins of a metal found close together.

ore: a rock containing enough of a useful substance to make mining it worthwhile.

vein: a mineral deposit different from, and usually cutting across, the surrounding rocks. Most mineral and metal-bearing veins are deposits filling fractures. The veins were filled by hot, mineral-rich waters rising upwards from liquid volcanic magma. They are important sources of many metals, such as silver and gold, and also minerals such as gemstones. Veins are usually narrow, and were best suited to hand-mining. They are less exploited in the modern machine age.

EQUATION: The use of cyanide to extract silver

Silver cyanide complex + zinc ⇨ zinc cyanide complex + silver

$$2NaAg(CN)_2(aq) \ + \ Zn(s) \ \Rightarrow \ Na_2Zn(CN)_4(aq) \ + \ 2Ag(s)$$

Also...

The Comstock Lode was a zone nearly six kilometres long that included deposits of native silver. It was discovered in the Sierra Nevada mountains of Nevada, USA, in 1859. It was named after the prospector who found the lode, Henry T. P. Comstock.

The area produced one of the richest silver rushes in history, probably yielding some $300 million at 19th century values (many billions in today's values). Virginia City was founded and grew up around the lode. When the silver was mined out, people moved away, and Virginia City is now a ghost town.

◀ The wealth of the town of Taxco in central Mexico was based on a silver lode discovered nearby. The lode is still worked, although most of the native metal has been removed in the past two centuries. The grand church and other buildings reflect the money that the exploitation of silver brought to this otherwise poor rural area.

Silver in photography

Because silver compounds are light-sensitive, about four-tenths of all the silver used industrially goes in to making photographic film.

The film is made of a plastic base over which is spread a thin layer of gelatin that contains silver salt. This gelatin layer is known as the emulsion layer.

Various silver salts are used in the emulsion. Silver iodide reacts fastest to light and is used for fast-speed films. Silver bromide is slightly less sensitive and is used for slower speeds. Silver bromide and silver chloride are light-sensitive chemicals that are placed on the surface of developing papers.

How light-sensitive silver salts work

Silver salts become reduced upon exposure to light. This converts parts of the salts to silver metal.

When the film is placed in the developer, another chemical reaction takes place, in which the silver salts are further reduced. Those that were not affected by light are reduced more slowly, leaving some areas darker than others.

The developed film is placed in a "fixing" solution to stop the process after it has fully reacted with the silver salts that were exposed to light. The process leaves a negative image on the acetate film.

A print is made by shining light through the negative on to photographic paper that has a coating containing silver chloride crystals. This material has to be developed and fixed in the same way as the film. At the end of the process a positive print is obtained.

▼▼ These two test tubes show the effect of light on some silver compounds. The test tube above shows a precipitate of silver chloride the instant after it was produced. The one below shows the change that has occurred within one minute. The way that silver salts darken on exposure to light is the basis of the photographic process.

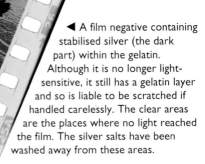

◄ A film negative containing stabilised silver (the dark part) within the gelatin. Although it is no longer light-sensitive, it still has a gelatin layer and so is liable to be scratched if handled carelessly. The clear areas are the places where no light reached the film. The silver salts have been washed away from these areas.

Also...

Silver is not attacked by most acids. However, it does react with concentrated nitric acid, liberating brown nitrogen dioxide gas and forming a solution of silver nitrate. This solution will precipitate halides. Silver chloride is a white solid that quickly darkens in sunlight; silver bromide is a pale yellow solid, and silver iodide is a dark yellow solid. All three of these solids disappear into a colourless solution if sodium thiosulphate solution (hypo) is added. This is why "hypo-fixer" works, since this solution can be washed off the film or photographic print paper.

▲ A positive print from the last century. The yellow colouring on this print results from the same process as the tarnishing described on page 29. The hydrogen sulphide gas present in the air reacts with the silver in the print, darkening it.

▲ Sodium thiosulphate

EQUATION: Silver bromide and photographers' hypo

Sodium thiosulphate (photographers' hypo) + silver bromide ⇨ silver complex + sodium bromide

$$2Na_2S_2O_3(aq) \quad + \quad AgBr(s) \quad \Rightarrow \quad Na_3Ag(S_2O_3)_2(aq) \quad + \quad NaBr(aq)$$

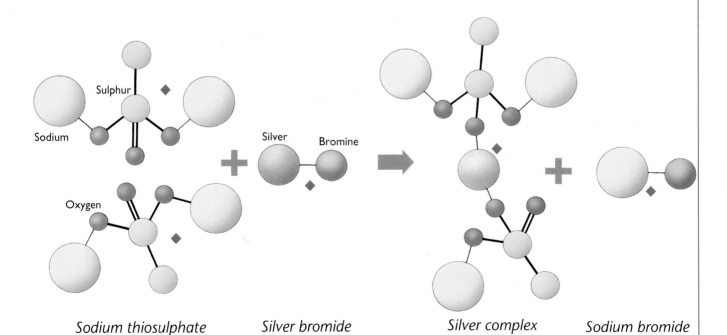

Sodium thiosulphate   Silver bromide   Silver complex   Sodium bromide

33

Everyday uses of silver

One of the most common objects around the home is the mirror. Mirrors reflect light rays to form an image.

The earliest form of mirror was a disk of polished bronze. However, although this gave some kind of image, the poor reflecting properties of bronze, and the fact that it tarnishes quickly, were serious drawbacks.

The next improvement, and in common use in the Middle Ages, was to use sheets of glass with metal foil attached to the back. Silver was the preferred metal for this because it is soft and could be beaten into thin sheets, and because it is highly reflective.

However, silver tarnishes over time in polluted air (the surface develops a dark silver sulphide coating). To produce a durable mirror, it was necessary to combine the strength of glass and the reflecting properties of silver by binding the silver to the glass so that hydrogen sulphide in the polluted air could not tarnish the surface.

In 1835 Justus von Liebig developed a way of depositing silver onto glass. The method, known as silvering, is still in use today.

Making a mirror in a tube

To make a mirror in a test tube, silver nitrate, silver hydroxide and ammonia are mixed together in solution together with a reducing agent, in this case glucose.
The tube is then placed in a bath of hot water. Silver is precipitated onto the warm glass tube.

▶ A silvered mirror produced on the inside of a test tube.

High-quality mirrors

Specialised mirrors are needed for precision scientific work. For these, silver is vaporised by heating it in a vacuum chamber. It is then condensed on to the glass as a thin even coating.

For a large-scale use of mirrors, such as in solar power stations, aluminium is often used as a substitute for silver. It is not quite such a good reflector, but it is very much cheaper.

amalgam: a liquid alloy of mercury with another metal.

cell: a vessel containing two electrodes and an electrolyte.

electrolyte: a solution that conducts electricity.

◄▲ The metal-based fillings in grinding teeth are made with either silver or gold amalgams.

In silver-based amalgam the proportions of metals by weight are: 52% mercury, 33% silver, 12.5% tin, 2% copper and 0.5% zinc.

As the amalgam sets, it expands slightly, locking itself into the tooth cavity.

Also...

If you have metal-based amalgam fillings and you chew a piece of "silver" or other metal foil, it is possible to get a shooting sensation through the filled teeth. In effect you will be repeating one of the world's earliest electrical experiments, where Luigi Galvani made frogs' legs twitch by using the body fluids and two different metals to create a battery (an electrical cell). In the case of the silver foil, the amalgam and the silver foil are two different metal electrodes, and your saliva is the electrolyte of the cell, so that a current will flow. This current may well cause a painful sensation in your tooth if the filled cavity is deep and the filling is close to the dental nerve.

Gold in the Earth

Gold mostly forms in veins in rocks. The source of gold is probably hot liquids that boiled off the great molten bodies that feed volcanoes. These hot fluids forced their way up through the cracks in the overlying rocks, where they cooled and solidified.

Gold is among a number of metals (including silver, copper and tin) that form in this way. Scientists call the veins hydrothermal (hot water) deposits.

Gold is so unreactive chemically that it does not easily form oxides or other compounds and remains as "native" (pure) metal in the rocks.

Mining

Vein, or lode, mining is the most important of all gold recovery methods (see page 30). Although each ounce of gold recovered requires about 100,000 ounces of ore to be processed, so much gold is deposited in rock veins that this method accounts for more than half of the world's total gold production today. The gold in the veins may be microscopic particles, nuggets, sheets, or gold compounds. Regardless of how it is found, the ore requires extensive extraction and refining.

◄ A vein carrying gold in a mine tunnel.

▼ A modern open-cast gold mine near Bendigo, Australia.

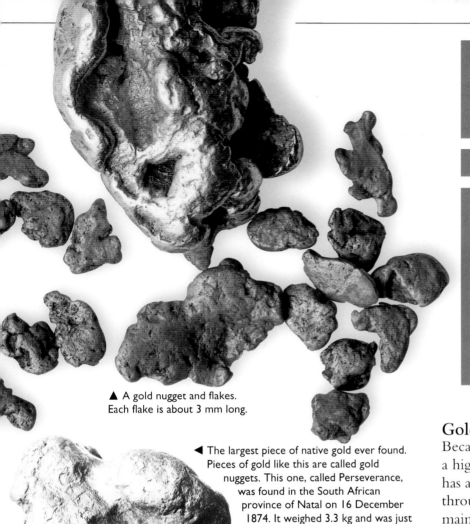

hydrothermal: a process in which hot water is involved. It is usually used in the context of rock formation because hot water and other fluids sent outwards from liquid magmas are important carriers of metals and the minerals that form gemstones.

lode: a deposit in which a number of veins of a metal found close together.

vein: a mineral deposit different from, and usually cutting across, the surrounding rocks. Most mineral and metal-bearing veins are deposits filling fractures. The veins were filled by hot, mineral-rich waters rising upwards from liquid volcanic magma. They are important sources of many metals, such as silver and gold, and also minerals such as gemstones. Veins are usually narrow, and were best suited to hand-mining. They are less exploited in the modern machine age.

▲ A gold nugget and flakes. Each flake is about 3 mm long.

◄ The largest piece of native gold ever found. Pieces of gold like this are called gold nuggets. This one, called Perseverance, was found in the South African province of Natal on 16 December 1874. It weighed 3.3 kg and was just over 12 cm long.

▼ Early 20th-century mining had become big business, rather than the province of many small-scale prospectors working by hand.

Gold rushes

Because people have put such a high value on gold, its presence has attracted much attention throughout the ages. One of the main reasons for Spanish and Portuguese explorers taking an interest in South America was to plunder them as a source of gold.

The 19th century saw several major gold rushes in the United States, Canada, Russia, South Africa and Australia. The 1848–49 gold rush in California produced more new gold than had been found in the previous three centuries and made the United States the largest gold producer in the world. But as these gold fields were worked out, others took their place in Alaska, Australia (particularly Victoria) and South Africa (The Transvaal). Of the probable gold reserves, about half are to be found in the Witwatersrand area of South Africa.

Gold at the surface

Much of the gold found during a gold rush was collected from river beds rather than from veins of gold dug from mines. Gold occurs in river beds as a result of natural landscape erosion. Where the gold veins are exposed at the land surface, erosion eventually breaks them down into small fragments and carries them away in streams and rivers.

However, because gold is such a heavy element, it is not easily moved and so it accumulates in the river bed. The fragments of gold are very small, but they are not found among the tiny fragments of other kinds of rock. Rather, they settle out among gravels and larger pieces of rock. This is because sand and pebbles are made of silica, a mineral that is much less dense than gold. A fast-flowing stream will carry small, but denser, particles of heavy gold along with larger, but less dense gravel.

Sluice boxes and dredging

The sluice box, shown below, is a much more efficient way of collecting gold. The larger pebbles are screened away and the other fragments washed down a chute. Bars placed along the chute collect the heavy gold, while the rest is washed away. Washing river sediment over wool fleeces is another way to collect gold.

Most placer deposits, especially those worked in navigable rivers and in coastal areas, are now mined on a far larger scale using dredgers.

▼ A man using a sluice box in Canada at the turn of the century.

38

Panning

The accumulation of gold in river sediments is known as a placer deposit. The gold can be separated out by hand in a method known as gold panning. Because it is heavy, gold resists being washed away. The gold particles can be separated from sand by swirling the pan around in a little water. The gold will remain in the centre of the pan while the sand swirls off the edges.

amalgam: a liquid alloy of mercury with another metal.

placer deposit: a kind of ore body made of a sediment that contains fragments of gold ore eroded from a mother lode and transported by rivers and/or ocean currents.

sediment: material that settles out at the bottom of a liquid when it is still.

◀ Panning gold in a river in Alaska.

Amalgamation: using mercury to refine gold

Another way to recover gold from placer deposits is by making the gold into a liquid alloy with mercury, known as an amalgam.

In this age-old process, the mercury and gold-containing sediment are heated, causing the gold to amalgamate with the mercury. The amalgam can then be drawn off and dissolved in dilute sodium cyanide. When zinc is added to the solution, gold is precipitated to the bottom of the vessel.

Amalgamation recovers about two-thirds of the gold in an ore. To extract the remainder, other chemical methods have to be used.

This process uses a range of highly poisonous substances and has to be carried out in carefully controlled conditions. Nevertheless, despite the risks, refining using mercury is done on a small scale by prospectors, for example, in the depths of the Brazilian rainforest. The waste products of this process are flushed into the rivers, where they pollute whole stretches of river water that are used as drinking supplies.

Many of the prospectors breathe in the mercury vapour as the amalgam is heated, causing long-term (and even fatal) damage to their health.

Chemical refining

Chemical refining is the last stage of the refining process. It can be applied to gold sediment that has already been partly refined using amalgamation or to the sludge produced in copper and zinc refining.

Like silver, gold can be dissolved in sodium cyanide. When metallic zinc is added to the solution, the gold is precipitated.

EQUATION: Precipitation of gold using sodium cyanide and zinc

Sodium cyanide + gold amalgam + water + oxygen ⇨ cyanide complex + sodium hydroxide

$$8NaCN(s) + 4Au(s) + 2H_2O(l) + O_2(g) \Rightarrow 4NaAu(CN)_2(aq) + 4NaOH(aq)$$

Cyanide complex + zinc ⇨ cyanide complex + gold

$$2NaAu(CN)_2(aq) + Zn(s) \Rightarrow Na_2Zn(CN)_4(aq) + 2Au(s)$$

Uses of gold

Gold is the most easily worked of all metals. This means that it can be drawn out into fine wires or beaten into sheets so thin that they are almost transparent. At this stage the gold is just a few atoms thick.

One of the most common uses of thin sheets of gold is as gold leaf, a decorative material that is used on furniture, walls and religious objects. Many people use gold leaf as part of their religious devotions. Gold has also played an important part in the world's money supply (see next page).

Carat

The purity of gold is measured in carats. Pure gold is known as 24 carat gold. Other forms of gold are alloys of gold with silver, copper or nickel.

The carat number shows how much gold is in each alloy. Thus 8 carat gold contains $^8/_{24}$ths or one-third gold; 18 carat gold contains $^{18}/_{24}$ths or three-quarters gold and so on. White gold has the lowest concentration of gold, being $^6/_{24}$ths gold and one-quarter nickel.

▲ Gold has been used decoratively since the earliest times both for jewellery and decorative ware, and for adornment to buildings (as in St. Mark's cathedral, Venice, shown here). Gold retains its shine even when used outdoors, which is why gold leaf is used on the domes of some United States capitol buildings.

▼ Gold leaf being used in a Buddhist ceremony.

The reactivity of gold

Gold is near the bottom of the reactivity series because it almost unreactive. Gold will not react with the air to form oxides. Traditionally only a mixture of concentrated nitric acid and concentrated hydrochloric acid (called *aqua regia*) would dissolve gold.

REACTIVITY SERIES	
Element	*Reactivity*
potassium	*most reactive*
sodium	
calcium	
magnesium	
aluminium	
manganese	
chromium	
zinc	
iron	
cadmium	
tin	
lead	
copper	
mercury	
silver	
gold	
platinum	*least reactive*

electrolysis: an electrical–chemical process that uses an electric current to cause the break up of a compound and the movement of metal ions in a solution. The process happens in many natural situations (as for example in rusting) and is also commonly used in industry for purifying (refining) metals or for plating metal objects with a fine, even metal coating.

ion: an atom, or group of atoms, that has gained or lost one or more electrons and so developed an electrical charge.

reactivity: the tendency of a substance to react with other substances. The term is most widely used in comparing the reactivity of metals. Metals are arranged in a reactivity series.

Gold and coinage

The first use of metal as a form of currency goes back to the ancient Egyptians about six thousand years ago. Gold, silver and copper were most often used, hence their grouping into the coinage metals. The metals were poured into moulds to produce blank discs called planchets. Markings were stamped on their faces by pressing them between two bronze dies. The coins usually had their value on one side and the head of the ruler on the other as a guarantee.

Gold will not stand up to handling unless it is alloyed with other metals. For example, in the United States gold coins (made until 1933) were 90% gold and the remainder silver and copper.

Gold was once used as a worldwide currency reserve, and was held in vaults as a guarantee of the paper money in circulation. This system no longer operates, and gold can be bought and sold on the open market. However, it is still true that when people feel nervous about a currency they buy gold, thus forcing up gold prices.

There are still large reserves of gold, the largest being at Fort Knox, Kentucky, USA, which contains both bullion (bars) and coins.

▼ A Krugerrand, one of the world's most famous gold coins.

Gold alloys and gold plating

Gold is not only rare, it is a very soft metal. For many reasons it is sensible to alloy it with other metals to improve its characteristics for use.

Gold is most often alloyed with copper and silver. This changes the strength and colour of the alloy as well as making the gold go farther. The resulting alloys are cheaper than pure gold.

"Green gold" is an alloy of gold, silver and copper with silver in the greatest proportion. In "red gold" copper dominates the alloy, and an alloy of gold and nickel is called "white gold".

Gold can be formed into alloys with mercury called amalgams; these are often used as dental fillings. Gold can also be plated on to objects to increase their resistance to corrosion.

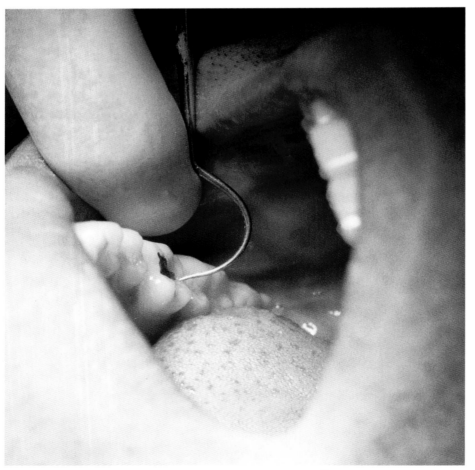

▲ The brooch shown here is a real orchid that has been coated in gold using electrolysis. The orchid was dipped in a resin, and the edges of the petals painted with a metal paint so that it could be made into an electrical conductor. Then it was suspended in a bath of electrolyte. As an electric current flowed, the gold was deposited on to the metal-coated flower.

◄ A gold filling in a lower tooth. In some countries gold fillings are a sign of wealth. Gold is a suitable material for a filling because it will amalgamate with mercury and because it does not react with body fluids, neither releasing toxic substances nor being corroded by the body.

The metal-based fillings in grinding teeth are made with either silver or gold amalgams. Gold fillings contain gold alloyed with palladium and copper. Rhodium is often added as a hardener for the gold–palladium alloy. Gold instead of silver is also often used in places where the filling is more liable to show, such as in holes in the front teeth.

Using gold in industry

Gold is almost entirely nonreactive. Although this means that gold will not form compounds, it also means that, where nonreactive materials are needed, gold has an important part to play. Because it is rare, it is also expensive to use. This is why gold is used only where such materials are important.

Gold is a good conductor of electricity and does not oxidise, so it is important for use in electrical circuits, such as microelectronics, and for connectors and switches. Most of the gold used is electroplated on to some less expensive base material.

Also, because it doesn't corrode, gold is used in places where there are corrosive atmospheres such as in certain chemical processes. For the same reason, gold can be used safely in amalgams to make dental fillings.

Gold can also have an important role in tiny amounts. For example it is used in glass to produce a heat shield.

alloy: a mixture of a metal and various other elements.

electrolysis: an electrical–chemical process that uses an electric current to cause the break up of a compound and the movement of metal ions in a solution. The process happens in many natural situations (as for example in rusting) and is also commonly used in industry for purifying (refining) metals or for plating metal objects with a fine, even metal coating.

electrolyte: a solution that conducts electricity.

electroplating: depositing a thin layer of a metal onto the surface of another substance using electrolysis.

Gold plating

Gold can be deposited as a fine layer on the surface of many materials. Usually it is applied by electrolysis after being dissolved in a cyanide solution. The thickness of gold plate obtained this way can be as little as two-millionths of a centimetre.

Some metals become coated with gold automatically if they are dipped in a gold cyanide solution, because a natural battery is established between the gold and the metal. Gold can be deposited on to nickel in this way.

◀ Gold does not oxidise, so no insulating oxide layer builds up on its surface (as happens, for example, with aluminium). This is the reason gold plated contacts are used in many electrical circuits where it is important that good long-term contact is made. This is a terminator from a computer.

Key facts about...

Copper

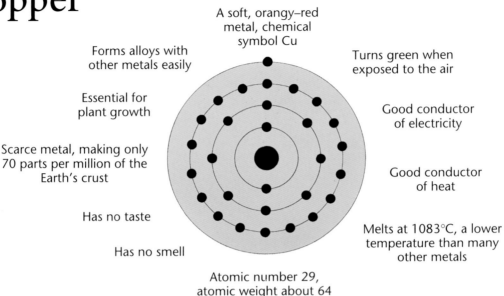

Forms alloys with other metals easily

A soft, orangy–red metal, chemical symbol Cu

Turns green when exposed to the air

Essential for plant growth

Good conductor of electricity

Scarce metal, making only 70 parts per million of the Earth's crust

Good conductor of heat

Has no taste

Has no smell

Melts at 1083°C, a lower temperature than many other metals

Atomic number 29, atomic weight about 64

Silver

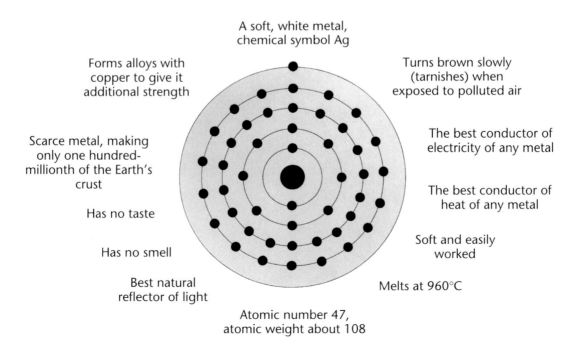

Forms alloys with copper to give it additional strength

A soft, white metal, chemical symbol Ag

Turns brown slowly (tarnishes) when exposed to polluted air

Scarce metal, making only one hundred-millionth of the Earth's crust

The best conductor of electricity of any metal

Has no taste

The best conductor of heat of any metal

Has no smell

Soft and easily worked

Best natural reflector of light

Melts at 960°C

Atomic number 47, atomic weight about 108

Gold

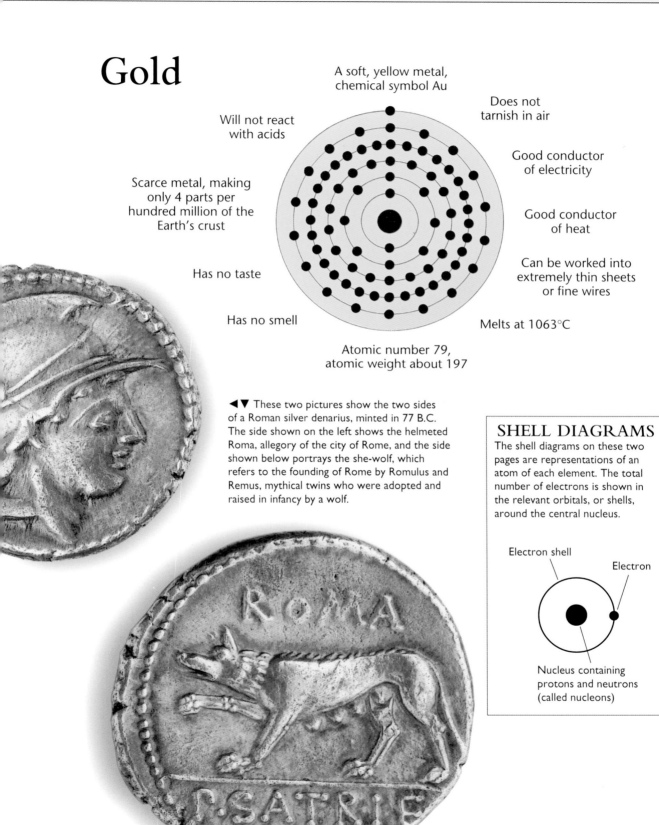

A soft, yellow metal, chemical symbol Au

Does not tarnish in air

Will not react with acids

Good conductor of electricity

Scarce metal, making only 4 parts per hundred million of the Earth's crust

Good conductor of heat

Can be worked into extremely thin sheets or fine wires

Has no taste

Melts at 1063°C

Has no smell

Atomic number 79, atomic weight about 197

◄▼ These two pictures show the two sides of a Roman silver denarius, minted in 77 B.C. The side shown on the left shows the helmeted Roma, allegory of the city of Rome, and the side shown below portrays the she-wolf, which refers to the founding of Rome by Romulus and Remus, mythical twins who were adopted and raised in infancy by a wolf.

SHELL DIAGRAMS

The shell diagrams on these two pages are representations of an atom of each element. The total number of electrons is shown in the relevant orbitals, or shells, around the central nucleus.

Electron shell

Electron

Nucleus containing protons and neutrons (called nucleons)

The Periodic Table

The Periodic Table sets out the relationships among the elements of the Universe. According to the Periodic Table, certain elements fall into groups. The pattern of these groups has, in the past, allowed scientists to predict elements that had not at that time been discovered. It can still be used today to predict the properties of unfamiliar elements.

The Periodic Table was first described by a Russian teacher, Dmitry Ivanovich Mendeleev, between 1869 and 1870. He was interested in writing a chemistry textbook, and wanted to show his students that there were certain patterns in the elements that had been discovered. So he set out the elements (of which there were 57 at the time) according to their known properties. On the assumption that there was pattern to the elements, he left blank spaces where elements seemed to be missing. Using this first version of the Periodic Table, he was able to predict in detail the chemical and physical properties of elements that had not yet been discovered. Other scientists began to look for the missing elements, and they soon found them.

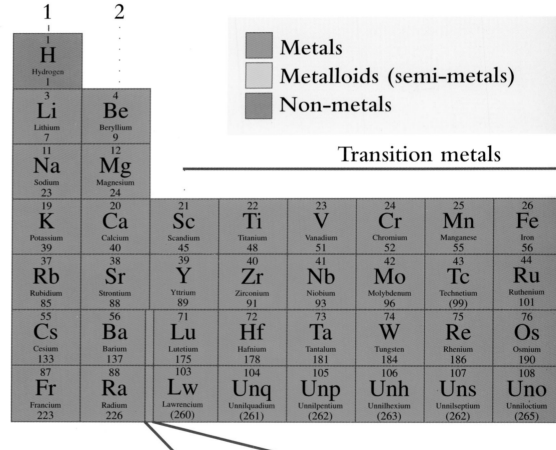

GROUP 1 2

Metals
Metalloids (semi-metals)
Non-metals

Transition metals

1 H Hydrogen 1							
3 Li Lithium 7	4 Be Beryllium 9						
11 Na Sodium 23	12 Mg Magnesium 24						
19 K Potassium 39	20 Ca Calcium 40	21 Sc Scandium 45	22 Ti Titanium 48	23 V Vanadium 51	24 Cr Chromium 52	25 Mn Manganese 55	26 Fe Iron 56
37 Rb Rubidium 85	38 Sr Strontium 88	39 Y Yttrium 89	40 Zr Zirconium 91	41 Nb Niobium 93	42 Mo Molybdenum 96	43 Tc Technetium (99)	44 Ru Ruthenium 101
55 Cs Cesium 133	56 Ba Barium 137	71 Lu Lutetium 175	72 Hf Hafnium 178	73 Ta Tantalum 181	74 W Tungsten 184	75 Re Rhenium 186	76 Os Osmium 190
87 Fr Francium 223	88 Ra Radium 226	103 Lw Lawrencium (260)	104 Unq Unnilquadium (261)	105 Unp Unnilpentium (262)	106 Unh Unnilhexium (263)	107 Uns Unnilseptium (262)	108 Uno Unniloctium (265)

Lanthanide metals

Actinoid metals

| 57 La Lanthanum 139 | 58 Ce Cerium 140 | 59 Pr Praseodymium 141 | 60 Nd Neodymium 144 |
| 89 Ac Actinium (227) | 90 Th Thorium 232 | 91 Pa Protactinium 231 | 92 U Uranium 238 |

Hydrogen did not seem to fit into the table, so he placed it in a box on its own. Otherwise the elements were all placed horizontally. When an element was reached with properties similar to the first one in the top row, a second row was started. By following this rule, similarities among the elements can be found by reading up and down. By reading across the rows, the elements progressively increase their atomic number. This number indicates the number of positively charged particles (protons) in the nucleus of each atom. This is also the number of negatively charged particles (electrons) in the atom.

The chemical properties of an element depend on the number of electrons in the outermost shell.

Atoms can form compounds by sharing electrons in their outermost shells. This explains why atoms with a full set of electrons (like helium, an inert gas) are unreactive, whereas atoms with an incomplete electron shell (such as chlorine) are very reactive. Elements can also combine by the complete transfer of electrons from metals to non-metals and the compounds formed contain ions.

Radioactive elements lose particles from their nucleus and electrons from their surrounding shells. As a result their atomic number changes and they become new elements.

Legend:
- Atomic (proton) number: 13
- Symbol: Al
- Name: Aluminium
- Approximate relative atomic mass (Approximate atomic weight): 27

3	4	5	6	7	0
					2 **He** Helium 4
5 **B** Boron 11	6 **C** Carbon 12	7 **N** Nitrogen 14	8 **O** Oxygen 16	9 **F** Fluorine 19	10 **Ne** Neon 20
13 **Al** Aluminium 27	14 **Si** Silicon 28	15 **P** Phosphorus 31	16 **S** Sulphur 32	17 **Cl** Chlorine 35	18 **Ar** Argon 40

				3	4	5	6	7	0
27 **Co** Cobalt 59	28 **Ni** Nickel 59	29 **Cu** Copper 64	30 **Zn** Zinc 65	31 **Ga** Gallium 70	32 **Ge** Germanium 73	33 **As** Arsenic 75	34 **Se** Selenium 79	35 **Br** Bromine 80	36 **Kr** Krypton 84
45 **Rh** Rhodium 103	46 **Pd** Palladium 106	47 **Ag** Silver 108	48 **Cd** Cadmium 112	49 **In** Indium 115	50 **Sn** Tin 119	51 **Sb** Antimony 122	52 **Te** Tellurium 128	53 **I** Iodine 127	54 **Xe** Xenon 131
77 **Ir** Iridium 192	78 **Pt** Platinum 195	79 **Au** Gold 197	80 **Hg** Mercury 201	81 **Tl** Thallium 204	82 **Pb** Lead 207	83 **Bi** Bismuth 209	84 **Po** Polonium (209)	85 **At** Astatine (210)	86 **Rn** Radon (222)
109 **Une** Unnilennium (266)									

61 **Pm** Promethium (145)	62 **Sm** Samarium 150	63 **Eu** Europium 152	64 **Gd** Gadolinium 157	65 **Tb** Terbium 159	66 **Dy** Dysprosium 163	67 **Ho** Holmium 165	68 **Er** Erbium 167	69 **Tm** Thulium 169	70 **Yb** Ytterbium 173
93 **Np** Neptunium (237)	94 **Pu** Plutonium (244)	95 **Am** Americium (243)	96 **Cm** Curium (247)	97 **Bk** Berkelium (247)	98 **Cf** Californium (251)	99 **Es** Einsteinium (252)	100 **Fm** Fermium (257)	101 **Md** Mendelevium (258)	102 **No** Nobelium (259)

Understanding equations

As you read through this book, you will notice that many pages contain equations using symbols. If you are not familiar with these symbols, read this page. Symbols make it easy for chemists to write out the reactions that are occurring in a way that allows a better understanding of the processes involved.

Symbols for the elements

The basis of the modern use of symbols for elements dates back to the 19th century. At this time a shorthand was developed using the first letter of the element wherever possible. Thus "O" stands for oxygen, "H" stands for hydrogen

and so on. However, if we were to use only the first letter, then there could be some confusion. For example, nitrogen and nickel would both use the symbols N. To overcome this problem, many elements are symbolised using the first two letters of their full name, and the second letter is lowercase. Thus although nitrogen is N, nickel becomes Ni. Not all symbols come from the English name; many use the Latin name instead. This is why, for example, gold is not G but Au (for the Latin *aurum*) and sodium has the symbol Na, from the Latin *natrium*.

Compounds of elements are made by combining letters. Thus the molecule carbon

Written and symbolic equations
In this book, important chemical equations are briefly stated in words (these are called word equations), and are then shown in their symbolic form along with the states.

What reaction the equation illustrates

EQUATION: The formation of calcium hydroxide

Word equation —

Calcium oxide + water ⇨ calcium hydroxide

Symbol equation —

$$CaO(s) \quad + \quad H_2O(l) \quad ⇨ \quad Ca(OH)_2(aq)$$

heated

Sometimes you will find additional descriptions below the symbolic equation.

Symbol showing the state:
s is for solid, l is for liquid,
g is for gas and aq is for aqueous.

Diagrams
Some of the equations are shown as graphic representations.

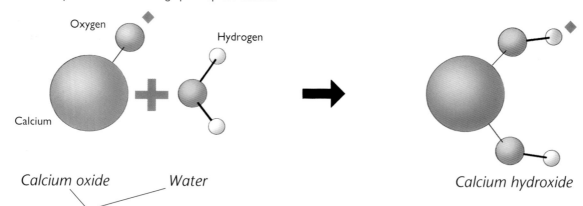

Oxygen

Hydrogen

Calcium

Calcium oxide   *Water*

Calcium hydroxide

Sometimes the written equation is broken up and put below the relevant stages in the graphic representation.

monoxide is CO. By using lowercase letters for the second letter of an element, it is possible to show that cobalt, symbol Co, is not the same as the molecule carbon monoxide, CO.

However, the letters can be made to do much more than this. In many molecules, atoms combine in unequal numbers. So, for example, carbon dioxide has one atom of carbon for every two of oxygen. This is shown by using the number 2 beside the oxygen, and the symbol becomes CO_2.

In practice, some groups of atoms combine as a unit with other substances. Thus, for example, calcium bicarbonate (one of the compounds used in some antacid pills) is written $Ca(HCO_3)_2$. This shows that the part of the substance inside the brackets reacts as a unit and the "2" outside the brackets shows the presence of two such units.

Some substances attract water molecules to themselves. To show this a dot is used. Thus the blue form of copper sulphate is written $CuSO_4.5H_2O$. In this case five molecules of water attract to one of copper sulphate.

When you see the dot, you know that this water can be driven off by heating; it is part of the crystal structure.

In a reaction substances change by rearranging the combinations of atoms. The way they change is shown by using the chemical symbols, placing those that will react (the starting materials, or reactants) on the left and the products of the reaction on the right. Between the two, chemists use an arrow to show which way the reaction is occurring.

It is possible to describe a reaction in words. This gives word equations, which are given throughout this book. However, it is easier to understand what is happening by using an equation containing symbols. These are also given in many places. They are not given when the equations are very complex.

In any equation both sides balance; that is, there must be an equal number of like atoms on both sides of the arrow. When you try to write down reactions, you, too, must balance your equation; you cannot have a few atoms left over at the end!

The symbols in brackets are abbreviations for the physical state of each substance taking part, so that (s) is used for solid, (l) for liquid, (g) for gas and (aq) for an aqueous solution, that is, a solution of a substance dissolved in water.

Atoms and ions

Each sphere represents a particle of an element. A particle can be an atom or an ion. Each atom or ion is associated with other atoms or ions through bonds – forces of attraction. The size of the particles and the nature of the bonds can be extremely important in determining the nature of the reaction or the properties of the compound.

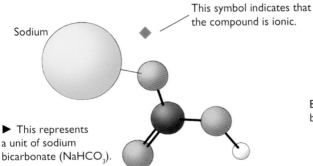

Sodium

This symbol indicates that the compound is ionic.

▶ This represents a unit of sodium bicarbonate ($NaHCO_3$).

The term "unit" is sometimes used to simplify the representation of a combination of ions.

Chemical symbols, equations and diagrams

The arrangement of any molecule or compound can be shown in one of the two ways shown below, depending on which gives the clearer picture. The left-hand diagram is called a ball-and-stick diagram because it uses rods and spheres to show the structure of the material. This example shows water, H_2O. There are two hydrogen atoms and one oxygen atom.

Bond shown by "stick"

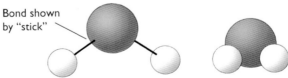

Colours too

The colours of each of the particles help differentiate the elements involved. The diagram can then be matched to the written and symbolic equation given with the diagram. In the case above, oxygen is red and hydrogen is grey.

Glossary of technical terms

absorb: to soak up a substance. Compare to adsorb.

acetone: a petroleum-based solvent.

acid: compounds containing hydrogen which can attack and dissolve many substances. Acids are described as weak or strong, dilute or concentrated, mineral or organic.

acidity: a general term for the strength of an acid in a solution.

acid rain: rain that is contaminated by acid gases such as sulphur dioxide and nitrogen oxides released by pollution.

adsorb/adsorption: to "collect" gas molecules or other particles on to the *surface* of a substance. They are not chemically combined and can be removed. (The process is called "adsorption".) Compare to absorb.

alchemy: the traditional "art" of working with chemicals that prevailed through the Middle Ages. One of the main challenges of alchemy was to make gold from lead. Alchemy faded away as scientific chemistry was developed in the 17th century.

alkali: a base in solution.

alkaline: the opposite of acidic. Alkalis are bases that dissolve, and alkaline materials are called basic materials. Solutions of alkalis have a pH greater than 7.0 because they contain relatively few hydrogen ions.

alloy: a mixture of a metal and various other elements.

alpha particle: a stable combination of two protons and two neutrons, which is ejected from the nucleus of a radioactive atom as it decays. An alpha particle is also the nucleus of the atom of helium. If it captures two electrons it can become a neutral helium atom.

amalgam: a liquid alloy of mercury with another metal.

amino acid: amino acids are organic compounds that are the building blocks for the proteins in the body.

amorphous: a solid in which the atoms are not arranged regularly (i.e. "glassy"). Compare with crystalline.

amphoteric: a metal that will react with both acids and alkalis.

anhydrous: a substance from which water has been removed by heating. Many hydrated salts are crystalline. When they are heated and the water is driven off, the material changes to an anhydrous powder.

anion: a negatively charged atom or group of atoms.

anode: the negative terminal of a battery or the positive electrode of an electrolysis cell.

anodising: a process that uses the effect of electrolysis to make a surface corrosion-resistant.

antacid: a common name for any compound that reacts with stomach acid to neutralise it.

antioxidant: a substance that prevents oxidation of some other substance.

aqueous: a solid dissolved in water. Usually used as "aqueous solution".

atom: the smallest particle of an element.

atomic number: the number of electrons or the number of protons in an atom.

atomised: broken up into a very fine mist. The term is used in connection with sprays and engine fuel systems.

aurora: the "northern lights" and "southern lights" that show as coloured bands of light in the night sky at high latitudes. They are associated with the way cosmic rays interact with oxygen and nitrogen in the air.

basalt: an igneous rock with a low proportion of silica (usually below 55%). It has microscopically small crystals.

base: a compound that may be soapy to the touch and that can react with an acid in water to form a salt and water.

battery: a series of electrochemical cells.

bauxite: an ore of aluminium, of which about half is aluminium oxide.

becquerel: a unit of radiation equal to one nuclear disintegration per second.

beta particle: a form of radiation in which electrons are emitted from an atom as the nucleus breaks down.

bleach: a substance that removes stains from materials either by oxidising or reducing the staining compound.

boiling point: the temperature at which a liquid boils, changing from a liquid to a gas.

bond: chemical bonding is either a transfer or sharing of electrons by two or more atoms. There are a number of types of chemical bond, some very strong (such as covalent bonds), others weak (such as hydrogen bonds). Chemical bonds form because the linked molecule is more stable than the unlinked atoms from which it formed. For example, the hydrogen molecule (H_2) is more stable than single atoms of hydrogen, which is why hydrogen gas is always found as molecules of two hydrogen atoms.

brass: a metal alloy principally of copper and zinc.

brazing: a form of soldering, in which brass is used as the joining metal.

brine: a solution of salt (sodium chloride) in water.

bronze: an alloy principally of copper and tin.

buffer: a chemistry term meaning a mixture of substances in solution that resists a change in the acidity or alkalinity of the solution.

capillary action: the tendency of a liquid to be sucked into small spaces, such as between objects and through narrow-pore tubes. The force to do this comes from surface tension.

catalyst: a substance that speeds up a chemical reaction but itself remains unaltered at the end of the reaction.

cathode: the positive terminal of a battery or the negative electrode of an electrolysis cell.

cathodic protection: the technique of making the object that is to be protected from corrosion into the cathode of a cell. For example, a material, such as steel, is protected by coupling it with a more reactive metal, such as magnesium. Steel forms the cathode and magnesium the anode. Zinc protects steel in the same way.

cation: a positively charged atom or group of atoms.

caustic: a substance that can cause burns if it touches the skin.

cell: a vessel containing two electrodes and an electrolyte that can act as an electrical conductor.

ceramic: a material based on clay minerals, which has been heated so that it has chemically hardened.

chalk: a pure form of calcium carbonate made of the crushed bodies of microscopic sea creatures, such as plankton and algae.

change of state: a change between one of the three states of matter, solid, liquid and gas.

chlorination: adding chlorine to a substance.

cladding: a surface sheet of material designed to protect other materials from corrosion.

clay: a microscopically small plate-like mineral that makes up the bulk of many soils. It has a sticky feel when wet.

combustion: the special case of oxidisation of a substance where a considerable amount of heat and usually light are given out. Combustion is often referred to as "burning".

compound: a chemical consisting of two or more elements chemically bonded together. Calcium atoms can combine with carbon atoms and oxygen atoms to make calcium carbonate, a compound of all three atoms.

condensation nuclei: microscopic particles of dust, salt and other materials suspended in the air, which attract water molecules.

conduction: (i) the exchange of heat (heat conduction) by contact with another object or (ii) allowing the flow of electrons (electrical conduction).

convection: the exchange of heat energy with the surroundings produced by the flow of a fluid due to being heated or cooled.

corrosion: the *slow* decay of a substance resulting from contact with gases and liquids in the environment. The term is often applied to metals. Rust is the corrosion of iron.

corrosive: a substance, either an acid or an alkali, that *rapidly* attacks a wide range of other substances.

cosmic rays: particles that fly through space and bombard all atoms on the Earth's surface. When they interact with the atmosphere they produce showers of secondary particles.

covalent bond: the most common form of strong chemical bonding, which occurs when two atoms *share* electrons.

cracking: breaking down complex molecules into simpler components. It is a term particularly used in oil refining.

crude oil: a chemical mixture of petroleum liquids. Crude oil forms the raw material for an oil refinery.

crystal: a substance that has grown freely so that it can develop external faces. Compare with crystalline, where the atoms are not free to form individual crystals and amorphous where the atoms are arranged irregularly.

crystalline: the organisation of atoms into a rigid "honeycomb-like" pattern without distinct crystal faces.

crystal systems: seven patterns or systems into which all of the world's crystals can be grouped. They are: cubic, hexagonal, rhombohedral, tetragonal, orthorhombic, monoclinic and triclinic.

cubic crystal system: groupings of crystals that look like cubes.

curie: a unit of radiation. The amount of radiation emitted by 1 g of radium each second. (The curie is equal to 37 billion becquerels.)

current: an electric current is produced by a flow of electrons through a conducting solid or ions through a conducting liquid.

decay (radioactive decay): the way that a radioactive element changes into another element because of loss of mass through radiation. For example uranium decays (changes) to lead.

decompose: to break down a substance (for example by heat or with the aid of a catalyst) into simpler components. In such a chemical reaction only one substance is involved.

dehydration: the removal of water from a substance by heating it, placing it in a dry atmosphere, or through the action of a drying agent.

density: the mass per unit volume (e.g. g/cc).

desertification: a process whereby a soil is allowed to become degraded to a state in which crops can no longer grow, i.e. desert-like. Chemical desertification is usually the result of contamination with halides because of poor irrigation practices.

detergent: a petroleum-based chemical that removes dirt.

diaphragm: a semipermeable membrane – a kind of ultra-fine mesh filter – that will allow only small ions to pass through. It is used in the electrolysis of brine.

diffusion: the slow mixing of one substance with another until the two substances are evenly mixed.

digestive tract: the system of the body that forms the pathway for food and its waste products. It begins at the mouth and includes the stomach and the intestines.

dilute acid: an acid whose concentration has been reduced by a large proportion of water.

diode: a semiconducting device that allows an electric current to flow in only one direction.

disinfectant: a chemical that kills bacteria and other microorganisms.

dissociate: to break apart. In the case of acids it means to break up forming hydrogen ions. This is an example of ionisation. Strong acids dissociate completely. Weak acids are not completely ionised and a solution of a weak acid has a relatively low concentration of hydrogen ions.

dissolve: to break down a substance in a solution without a resultant reaction.

distillation: the process of separating mixtures by condensing the vapours through cooling.

doping: adding metal atoms to a region of silicon to make it semiconducting.

dye: a coloured substance that will stick to another substance, so that both appear coloured.

electrode: a conductor that forms one terminal of a cell.

electrolysis: an electrical–chemical process that uses an electric current to cause the break up of a compound and the movement of metal ions in a solution. The process happens in many natural situations (as for example in rusting) and is also commonly used in industry for purifying (refining) metals or for plating metal objects with a fine, even metal coating.

electrolyte: a solution that conducts electricity.

electron: a tiny, negatively charged particle that is part of an atom. The flow of electrons through a solid material such as a wire produces an electric current.

electroplating: depositing a thin layer of a metal onto the surface of another substance using electrolysis.

element: a substance that cannot be decomposed into simpler substances by chemical means

emulsion: tiny droplets of one substance dispersed in another. A common oil in water emulsion is milk. The tiny droplets in an emulsion tend to come together, so another stabilising substance is often needed to wrap the particles of grease and oil in a stable coat. Soaps and detergents are such agents. Photographic film is an example of a solid emulsion.

endothermic reaction: a reaction that takes heat from the surroundings. The reaction of carbon monoxide with a metal oxide is an example.

enzyme: organic catalysts in the form of proteins in the body that speed up chemical reactions. Every living cell contains hundreds of enzymes, which ensure that the processes of life continue. Should enzymes be made inoperative, such as through mercury poisoning, then death follows.

ester: organic compounds, formed by the reaction of an alcohol with an acid, which often have a fruity taste.

evaporation: the change of state of a liquid to a gas. Evaporation happens below the boiling point and is used as a method of separating out the materials in a solution.

exothermic reaction: a reaction that gives heat to the surroundings. Many oxidation reactions, for example, give out heat.

explosive: a substance which, when a shock is applied to it, decomposes very rapidly, releasing a very large amount of heat and creating a large volume of gases as a shock wave.

extrusion: forming a shape by pushing it through a die. For example, toothpaste is extruded through the cap (die) of the toothpaste tube.

fallout: radioactive particles that reach the ground from radioactive materials in the atmosphere.

fat: semi-solid energy-rich compounds derived from plants or animals and which are made of carbon, hydrogen and oxygen. Scientists call these esters.

feldspar: a mineral consisting of sheets of aluminium silicate. This is the mineral from which the clay in soils is made.

fertile: able to provide the nutrients needed for unrestricted plant growth.

filtration: the separation of a liquid from a solid using a membrane with small holes.

fission: the breakdown of the structure of an atom, popularly called "splitting the atom" because the atom is split into approximately two other nuclei. This is different from, for example, the small change that happens when radioactivity is emitted.

fixation of nitrogen: the processes that natural organisms, such as bacteria, use to turn the nitrogen of the air into ammonium compounds.

fixing: making solid and liquid nitrogen-containing compounds from nitrogen gas. The compounds that are formed can be used as fertilisers.

fluid: able to flow; either a liquid or a gas.

fluorescent: a substance that gives out visible light when struck by invisible waves such as ultraviolet rays.

flux: a material used to make it easier for a liquid to flow. A flux dissolves metal oxides and so prevents a metal from oxidising while being heated.

foam: a substance that is sufficiently gelatinous to be able to contain bubbles of gas. The gas bulks up the substance, making it behave as though it were semi-rigid.

fossil fuels: hydrocarbon compounds that have been formed from buried plant and animal remains. High pressures and temperatures lasting over millions of years are required. The fossil fuels are coal, oil and natural gas.

fraction: a group of similar components of a mixture. In the petroleum industry the light fractions of crude oil are those with the smallest molecules, while the medium and heavy fractions have larger molecules.

free radical: a very reactive atom or group with a "spare" electron.

freezing point: the temperature at which a substance changes from a liquid to a solid. It is the same temperature as the melting point.

fuel: a concentrated form of chemical energy. The main sources of fuels (called fossil fuels because they were formed by geological processes) are coal, crude oil and natural gas. Products include methane, propane and gasoline. The fuel for stars and space vehicles is hydrogen.

fuel rods: rods of uranium or other radioactive material used as a fuel in nuclear power stations.

fuming: an unstable liquid that gives off a gas. Very concentrated acid solutions are often fuming solutions.

fungicide: any chemical that is designed to kill fungi and control the spread of fungal spores.

fusion: combining atoms to form a heavier atom.

galvanising: applying a thin zinc coating to protect another metal.

gamma rays: waves of radiation produced as the nucleus of a radioactive element rearranges itself into a tighter cluster of protons and neutrons. Gamma rays carry enough energy to damage living cells.

gangue: the unwanted material in an ore.

gas: a form of matter in which the molecules form no definite shape and are free to move about to fill any vessel they are put in.

gelatinous: a term meaning made with water. Because a gelatinous precipitate is mostly water, it is of a similar density to water and will float or lie suspended in the liquid.

gelling agent: a semi-solid jelly-like substance.

gemstone: a wide range of minerals valued by people, both as crystals (such as emerald) and as decorative stones (such as agate). There is no single chemical formula for a gemstone.

glass: a transparent silicate without any crystal growth. It has a glassy lustre and breaks with a curved fracture. Note that some minerals have all these features and are therefore natural glasses. Household glass is a synthetic silicate.

glucose: the most common of the natural sugars. It occurs as the polymer known as cellulose, the fibre in plants. Starch is also a form of glucose. The breakdown of glucose provides the energy that animals need for life.

granite: an igneous rock with a high proportion of silica (usually over 65%). It has well-developed large crystals. The largest pink, grey or white crystals are feldspar.

Greenhouse Effect: an increase of the global air temperature as a result of heat released from burning fossil fuels being absorbed by carbon dioxide in the atmosphere.

gypsum: the name for calcium sulphate. It is commonly found as Plaster of Paris and wallboards.

half-life: the time it takes for the radiation coming from a sample of a radioactive element to decrease by half.

halide: a salt of one of the halogens (fluorine, chlorine, bromine and iodine).

halite: the mineral made of sodium chloride.

halogen: one of a group of elements including chlorine, bromine, iodine and fluorine.

heat-producing: see exothermic reaction.

high explosive: a form of explosive that will only work when it receives a shock from another explosive. High explosives are much more powerful than ordinary explosives. Gunpowder is not a high explosive.

hydrate: a solid compound in crystalline form that contains molecular water. Hydrates commonly form when a solution of a soluble salt is evaporated. The water that forms part of a hydrate crystal is known as the "water of crystallization". It can usually be removed by heating, leaving an anhydrous salt.

hydration: the absorption of water by a substance. Hydrated materials are not "wet" but remain firm, apparently dry, solids. In some cases, hydration makes the substance change colour, in many other cases there is no colour change, simply a change in volume.

hydrocarbon: a compound in which only hydrogen and carbon atoms are present. Most fuels are hydrocarbons, as is the simple plastic polyethene (known as polythene).

hydrogen bond: a type of attractive force that holds one molecule to another. It is one of the weaker forms of intermolecular attractive force.

hydrothermal: a process in which hot water is involved. It is usually used in the context of rock formation because hot water and other fluids sent outwards from liquid magmas are important carriers of metals and the minerals that form gemstones.

igneous rock: a rock that has solidified from molten rock, either volcanic lava on the Earth's surface or magma deep underground. In either case the rock develops a network of interlocking crystals.

incendiary: a substance designed to cause burning.

indicator: a substance or mixture of substances that change colour with acidity or alkalinity.

inert: nonreactive.

infra-red radiation: a form of light radiation where the wavelength of the waves is slightly longer than visible light. Most heat radiation is in the infra-red band.

insoluble: a substance that will not dissolve.

ion: an atom, or group of atoms, that has gained or lost one or more electrons and so developed an electrical charge. Ions behave differently from electrically neutral atoms and molecules. They can move in an electric field, and they can also bind strongly to solvent molecules such as water. Positively charged ions are called cations; negatively charged ions are called anions. Ions carry electrical current through solutions.

ionic bond: the form of bonding that occurs between two ions when the ions have opposite charges. Sodium cations bond with chloride anions to form common salt (NaCl) when a salty solution is evaporated. Ionic bonds are strong bonds except in the presence of a solvent.

ionise: to break up neutral molecules into oppositely charged ions or to convert atoms into ions by the loss of electrons.

ionisation: a process that creates ions.

irrigation: the application of water to fields to help plants grow during times when natural rainfall is sparse.

isotope: atoms that have the same number of protons in their nucleus, but which have different masses; for example, carbon-12 and carbon-14.

latent heat: the amount of heat that is absorbed or released during the process of changing state between gas, liquid or solid. For example, heat is absorbed when a substance melts and it is released again when the substance solidifies.

latex: (the Latin word for "liquid") a suspension of small polymer particles in water. The rubber that flows from a rubber tree is a natural latex. Some synthetic polymers are made as latexes, allowing polymerisation to take place in water.

lava: the material that flows from a volcano.

limestone: a form of calcium carbonate rock that is often formed of lime mud. Most limestones are light grey and have abundant fossils.

liquid: a form of matter that has a fixed volume but no fixed shape.

lode: a deposit in which a number of veins of a metal found close together.

lustre: the shininess of a substance.

magma: the molten rock that forms a balloon-shaped chamber in the rock below a volcano. It is fed by rock moving upwards from below the crust.

marble: a form of limestone that has been "baked" while deep inside mountains. This has caused the limestone to melt and reform into small interlocking crystals, making marble harder than limestone.

mass: the amount of matter in an object. In everyday use, the word weight is often used to mean mass.

melting point: the temperature at which a substance changes state from a solid to a liquid. It is the same as freezing point.

membrane: a thin flexible sheet. A semipermeable membrane has microscopic holes of a size that will selectively allow some ions and molecules to pass through but hold others back. It thus acts as a kind of sieve.

meniscus: the curved surface of a liquid that forms when it rises in a small bore, or capillary tube. The meniscus is convex (bulges upwards) for mercury and is concave (sags downwards) for water.

metal: a substance with a lustre, the ability to conduct heat and electricity and which is not brittle.

metallic bonding: a kind of bonding in which atoms reside in a "sea" of mobile electrons. This type of bonding allows metals to be good conductors and means that they are not brittle

metamorphic rock: formed either from igneous or sedimentary rocks, by heat and or pressure. Metamorphic rocks form deep inside mountains during periods of mountain building. They result from the remelting of rocks during which process crystals are able to grow. Metamorphic rocks often show signs of banding and partial melting.

micronutrient: an element that the body requires in small amounts. Another term is trace element.

mineral: a solid substance made of just one element or chemical compound. Calcite is a mineral because it consists only of calcium carbonate, halite is a mineral because it contains only sodium chloride, quartz is a mineral because it consists of only silicon dioxide.

mineral acid: an acid that does not contain carbon and that attacks minerals. Hydrochloric, sulphuric and nitric acids are the main mineral acids.

mineral-laden: a solution close to saturation.

mixture: a material that can be separated out into two or more substances using physical means.

molecule: a group of two or more atoms held together by chemical bonds.

monoclinic system: a grouping of crystals that look like double-ended chisel blades.

monomer: a building block of a larger chain molecule ("mono" means one, "mer" means part).

mordant: any chemical that allows dyes to stick to other substances.

native metal: a pure form of a metal, not combined as a compound. Native metal is more common in poorly reactive elements than in those that are very reactive.

neutralisation: the reaction of acids and bases to produce a salt and water. The reaction causes hydrogen from the acid and hydroxide from the base to be changed to water. For example, hydrochloric acid reacts with sodium hydroxide to form common salt and water. The term is more generally used for any reaction where the pH changes towards 7.0, which is the pH of a neutral solution.

neutron: a particle inside the nucleus of an atom that is neutral and has no charge.

noncombustible: a substance that will not burn.

noble metal: silver, gold, platinum, and mercury. These are the least reactive metals.

nuclear energy: the heat energy produced as part of the changes that take place in the core, or nucleus, of an element's atoms.

nuclear reactions: reactions that occur in the core, or nucleus of an atom.

nutrients: soluble ions that are essential to life.

octane: one of the substances contained in fuel.

ore: a rock containing enough of a useful substance to make mining it worthwhile.

organic acid: an acid containing carbon and hydrogen.

organic substance: a substance that contains carbon.

osmosis: a process where molecules of a liquid solvent move through a membrane (filter) from a region of low concentration to a region of high concentration of solute.

oxidation: a reaction in which the oxidising agent removes electrons. (Note that oxidising agents do not have to contain oxygen.)

oxide: a compound that includes oxygen and one other element.

oxidise: the process of gaining oxygen. This can be part of a controlled chemical reaction, or it can be the result of exposing a substance to the air, where oxidation (a form of corrosion) will occur slowly, perhaps over months or years.

oxidising agent: a substance that removes electrons from another substance (and therefore is itself reduced).

ozone: a form of oxygen whose molecules contain three atoms of oxygen. Ozone is regarded as a beneficial gas when high in the atmosphere because it blocks ultraviolet rays. It is a harmful gas when breathed in, so low level ozone, which is produced as part of city smog, is regarded as a form of pollution. The ozone layer is the uppermost part of the stratosphere.

pan: the name given to a shallow pond of liquid. Pans are mainly used for separating solutions by evaporation.

patina: a surface coating that develops on metals and protects them from further corrosion.

percolate: to move slowly through the pores of a rock.

period: a row in the Periodic Table.

Periodic Table: a chart organising elements by atomic number and chemical properties into groups and periods.

pesticide: any chemical that is designed to control pests (unwanted organisms) that are harmful to plants or animals.

petroleum: a natural mixture of a range of gases, liquids and solids derived from the decomposed remains of plants and animals.

pH: a measure of the hydrogen ion concentration in a liquid. Neutral is pH 7.0; numbers greater than this are alkaline, smaller numbers are acidic.

phosphor: any material that glows when energized by ultraviolet or electron beams such as in fluorescent tubes and cathode ray tubes. Phosphors, such as phosphorus, emit light after the source of excitation is cut off. This is why they glow in the dark. By contrast, fluorescors, such as fluorite, emit light only while they are being excited by ultraviolet light or an electron beam.

photon: a parcel of light energy.

photosynthesis: the process by which plants use the energy of the Sun to make the compounds they need for life. In photosynthesis, six molecules of carbon dioxide from the air combine with six molecules of water, forming one molecule of glucose (sugar) and releasing six molecules of oxygen back into the atmosphere.

pigment: any solid material used to give a liquid a colour.

placer deposit: a kind of ore body made of a sediment that contains fragments of gold ore eroded from a mother lode and transported by rivers and/or ocean currents.

plastic (material): a carbon-based material consisting of long chains (polymers) of simple molecules. The word plastic is commonly restricted to synthetic polymers.

plastic (property): a material is plastic if it can be made to change shape easily. Plastic materials will remain in the new shape. (Compare with elastic, a property where a material goes back to its original shape.)

plating: adding a thin coat of one material to another to make it resistant to corrosion.

playa: a dried-up lake bed that is covered with salt deposits. From the Spanish word for beach.

poison gas: a form of gas that is used intentionally to produce widespread injury and death. (Many gases are poisonous, which is why many chemical reactions are performed in laboratory fume chambers, but they are a byproduct of a reaction and not intended to cause harm.)

polymer: a compound that is made of long chains by combining molecules (called monomers) as repeating units. ("Poly" means many, "mer" means part).

polymerisation: a chemical reaction in which large numbers of similar molecules arrange themselves into large molecules, usually long chains. This process usually happens when there is a suitable catalyst present. For example, ethene reacts to form polythene in the presence of certain catalysts.

porous: a material containing many small holes or cracks. Quite often the pores are connected, and liquids, such as water or oil, can move through them.

precious metal: silver, gold, platinum, iridium, and palladium. Each is prized for its rarity. This category is the equivalent of precious stones, or gemstones, for minerals.

precipitate: tiny solid particles formed as a result of a chemical reaction between two liquids or gases.

preservative: a substance that prevents the natural organic decay processes from occurring. Many substances can be used safely for this purpose, including sulphites and nitrogen gas.

product: a substance produced by a chemical reaction.

protein: molecules that help to build tissue and bone and therefore make new body cells. Proteins contain amino acids.

proton: a positively charged particle in the nucleus of an atom that balances out the charge of the surrounding electrons

pyrite: "mineral of fire". This name comes from the fact that pyrite (iron sulphide) will give off sparks if struck with a stone.

pyrometallurgy: refining a metal from its ore using heat. A blast furnace or smelter is the main equipment used.

radiation: the exchange of energy with the surroundings through the transmission of waves or particles of energy. Radiation is a form of energy transfer that can happen through space; no intervening medium is required (as would be the case for conduction and convection).

radioactive: a material that emits radiation or particles from the nucleus of its atoms.

radioactive decay: a change in a radioactive element due to loss of mass through radiation. For example uranium decays (changes) to lead.

radioisotope: a shortened version of the phrase radioactive isotope.

radiotracer: a radioactive isotope that is added to a stable, nonradioactive material in order to trace how it moves and its concentration.

reaction: the recombination of two substances using parts of each substance to produce new substances.

reactivity: the tendency of a substance to react with other substances. The term is most widely used in comparing the reactivity of metals. Metals are arranged in a reactivity series.

reagent: a starting material for a reaction.

recycling: the reuse of a material to save the time and energy required to extract new material from the Earth and to conserve non-renewable resources.

redox reaction: a reaction that involves reduction and oxidation.

reducing agent: a substance that gives electrons to another substance. Carbon monoxide is a reducing agent when passed over copper oxide, turning it to copper and producing carbon dioxide gas. Similarly, iron oxide is reduced to iron in a blast furnace. Sulphur dioxide is a reducing agent, used for bleaching bread.

reduction: the removal of oxygen from a substance. See also: oxidation.

refining: separating a mixture into the simpler substances of which it is made. In the case of a rock, it means the extraction of the metal that is mixed up in the rock. In the case of oil it means separating out the fractions of which it is made.

refractive index: the property of a transparent material that controls the angle at which total internal reflection will occur. The greater the refractive index, the more reflective the material will be.

resin: natural or synthetic polymers that can be moulded into solid objects or spun into thread.

rust: the corrosion of iron and steel.

saline: a solution in which most of the dissolved matter is sodium chloride (common salt).

salinisation: the concentration of salts, especially sodium chloride, in the upper layers of a soil due to poor methods of irrigation.

salts: compounds, often involving a metal, that are the reaction products of acids and bases. (Note "salt" is also the common word for sodium chloride, common salt or table salt.)

saponification: the term for a reaction between a fat and a base that produces a soap.

saturated: a state where a liquid can hold no more of a substance. If any more of the substance is added, it will not dissolve.

saturated solution: a solution that holds the maximum possible amount of dissolved material. The amount of material in solution varies with the temperature; cold solutions can hold less dissolved solid material than hot solutions. Gases are more soluble in cold liquids than hot liquids.

sediment: material that settles out at the bottom of a liquid when it is still.

semiconductor: a material of intermediate conductivity. Semiconductor devices often use silicon when they are made as part of diodes, transistors or integrated circuits.

semipermeable membrane: a thin (membrane) of material that acts as a fine sieve, allowing small molecules to pass, but holding large molecules back.

silicate: a compound containing silicon and oxygen (known as silica).

sintering: a process that happens at moderately high temperatures in some compounds. Grains begin to fuse together even through they do not melt. The most widespread example of sintering happens during the firing of clays to make ceramics.

slag: a mixture of substances that are waste products of a furnace. Most slags are composed mainly of silicates.

smelting: roasting a substance in order to extract the metal contained in it.

smog: a mixture of smoke and fog. The term is used to describe city fogs in which there is a large proportion of particulate matter (tiny pieces of carbon from exhausts) and also a high concentration of sulphur and nitrogen gases and probably ozone.

soldering: joining together two pieces of metal using solder, an alloy with a low melting point.

solid: a form of matter where a substance has a definite shape.

soluble: a substance that will readily dissolve in a solvent.

solute: the substance that dissolves in a solution (e.g. sodium chloride in salt water).

solution: a mixture of a liquid and at least one other substance (e.g. salt water). Mixtures can be separated out by physical means, for example by evaporation and cooling.

solvent: the main substance in a solution (e.g. water in salt water).

spontaneous combustion: the effect of a very reactive material beginning to oxidise very quickly and bursting into flame.

stable: able to exist without changing into another substance.

stratosphere: the part of the Earth's atmosphere that lies immediately above the region in which clouds form. It occurs between 12 and 50 km above the Earth's surface.

strong acid: an acid that has completely dissociated (ionised) in water. Mineral acids are strong acids.

sublimation: the change of a substance from solid to gas, or vica versa, without going through a liquid phase.

substance: a type of material, including mixtures.

sulphate: a compound that includes sulphur and oxygen, for example, calcium sulphate or gypsum.

sulphide: a sulphur compound that contains no oxygen.

sulphite: a sulphur compound that contains less oxygen than a sulphate.

surface tension: the force that operates on the surface of a liquid, which makes it act as though it were covered with an invisible elastic film.

suspension: tiny particles suspended in a liquid.

synthetic: does not occur naturally, but has to be manufactured.

tarnish: a coating that develops as a result of the reaction between a metal and substances in the air. The most common form of tarnishing is a very thin transparent oxide coating.

thermonuclear reactions: reactions that occur within atoms due to fusion, releasing an immensely concentrated amount of energy.

thermoplastic: a plastic that will soften, can repeatedly be moulded it into shape on heating and will set into the moulded shape as it cools.

thermoset: a plastic that will set into a moulded shape as it cools, but which cannot be made soft by reheating.

titration: a process of dripping one liquid into another in order to find out the amount needed to cause a neutral solution. An indicator is used to signal change.

toxic: poisonous enough to cause death.

translucent: almost transparent.

transmutation: the change of one element into another.

vapour: the gaseous form of a substance that is normally a liquid. For example, water vapour is the gaseous form of liquid water.

vein: a mineral deposit different from, and usually cutting across, the surrounding rocks. Most mineral and metal-bearing veins are deposits filling fractures. The veins were filled by hot, mineral-rich waters rising upwards from liquid volcanic magma. They are important sources of many metals, such as silver and gold, and also minerals such as gemstones. Veins are usually narrow, and were best suited to hand-mining. They are less exploited in the modern machine age.

viscous: slow moving, syrupy. A liquid that has a low viscosity is said to be mobile.

vitreous: glass-like.

volatile: readily forms a gas.

vulcanisation: forming cross-links between polymer chains to increase the strength of the whole polymer. Rubbers are vulcanised using sulphur when making tyres and other strong materials.

weak acid: an acid that has only partly dissociated (ionised) in water. Most organic acids are weak acids.

weather: a term used by Earth scientists and derived from "weathering", meaning to react with water and gases of the environment.

weathering: the slow natural processes that break down rocks and reduce them to small fragments either by mechanical or chemical means.

welding: fusing two pieces of metal together using heat.

X-rays: a form of very short wave radiation.

Index

RAINFOREST

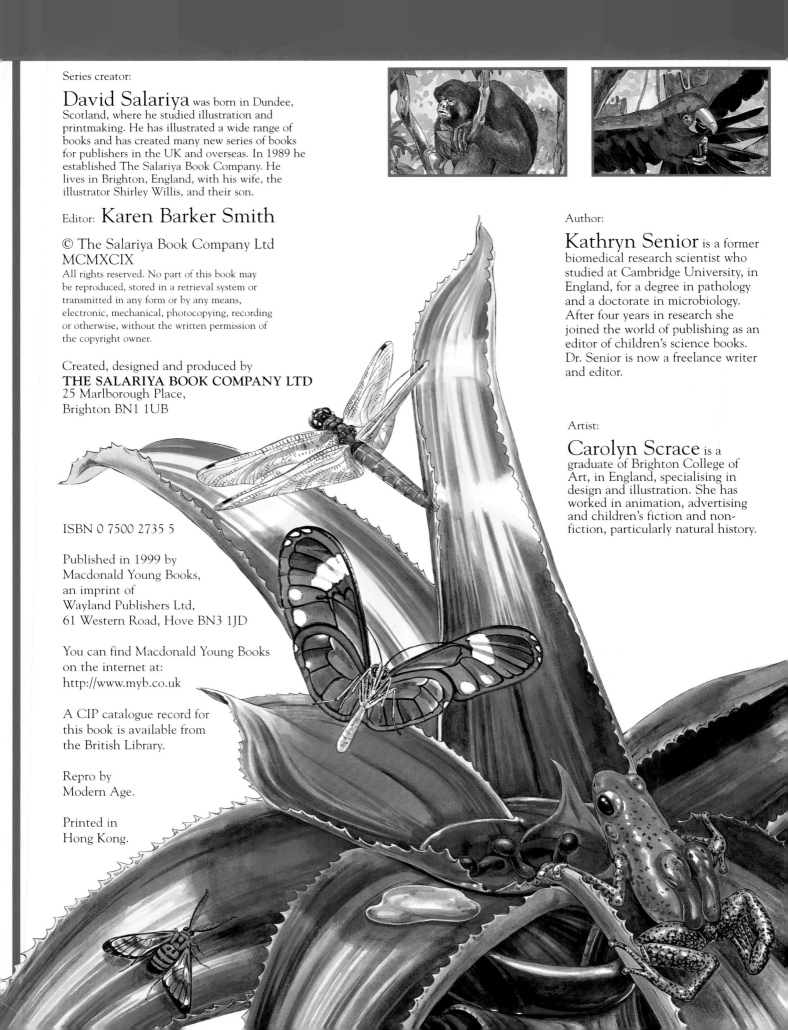

Series creator:

David Salariya was born in Dundee, Scotland, where he studied illustration and printmaking. He has illustrated a wide range of books and has created many new series of books for publishers in the UK and overseas. In 1989 he established The Salariya Book Company. He lives in Brighton, England, with his wife, the illustrator Shirley Willis, and their son.

Editor: **Karen Barker Smith**

Created, designed and produced by
THE SALARIYA BOOK COMPANY LTD
25 Marlborough Place,
Brighton BN1 1UB

ISBN 0 7500 2735 5

Published in 1999 by
Macdonald Young Books,
an imprint of
Wayland Publishers Ltd,
61 Western Road, Hove BN3 1JD

You can find Macdonald Young Books
on the internet at:
http://www.myb.co.uk

A CIP catalogue record for
this book is available from
the British Library.

Repro by
Modern Age.

Printed in
Hong Kong.

Author:

Kathryn Senior is a former biomedical research scientist who studied at Cambridge University, in England, for a degree in pathology and a doctorate in microbiology. After four years in research she joined the world of publishing as an editor of children's science books. Dr. Senior is now a freelance writer and editor.

Artist:

Carolyn Scrace is a graduate of Brighton College of Art, in England, specialising in design and illustration. She has worked in animation, advertising and children's fiction and non-fiction, particularly natural history.

FAST FORWARD

RAINFOREST

Written by
KATHRYN SENIOR

Illustrated by
CAROLYN SCRACE

Created and designed by
DAVID SALARIYA

MACDONALD YOUNG BOOKS

Contents

What is a Rainforest?

Rainforests are wet most of the time because of the 200 centimetres or more of rain that falls on them every year. They do exist in regions other than the tropics, but usually the term 'rainforest' refers to those found on or near the equator.

There are several different types of rainforest. Lowland rainforests thrive at low altitudes and are the most extensive. The canopy is about 45 metres above the ground and forms a dense covering with few gaps between the trees. Some taller trees jut out to heights of over 60 metres.

Several lowland rainforests are so near to rivers they are permanently flooded. Mangrove rainforests are those that have formed in such flooded areas.

Extracts from more than 5,000 plant and animal species found in rainforests are used in food, medicine and other products. At least 3,000 plant species that grow in the rainforests are known to contain compounds that can be used to develop new drugs to treat cancer. As the rainforests disappear, so do our chances of finding further useful substances in them.

The mangrove trees survive by growing roots that stick above the water.

Montane rainforests are found at higher altitudes, where it tends to be slightly colder. For this reason, the trees there do not grow as high – about 15–30 metres is the usual height of the canopy.

Before human beings existed, tropical rainforests of all types covered more than 20 million square kilometres of the planet. Over 16 million hectares of rainforest are destroyed every year when trees are cut down for timber, paper pulp and other wood products. By the end of the year 2000, there will be less than 8.8 million square kilometres of rainforest left. The areas of rainforest that still exist today are shown on the maps below.

Less than 5 percent of the world's tropical rainforests are in protected national parks. The other 95 percent can be plundered for valuable natural resources. As long as the destruction continues, one rainforest species becomes extinct every 15 minutes. If our use of the Earth's tropical rainforests does not change, they will all have disappeared by the year 2200. Approximately 25 percent of the planet's species will already be extinct by the middle of the 21st century.

Packed with Plants

The rainforest is a living, three-dimensional mosaic of plants. The canopy, the continuous cover provided by broad-leaved evergreen trees, towers as high as 60 metres above the ground. Under the thickest parts of the canopy few other plants are able grow. Where the top layer is less dense a rich ground level of ferns, herbs and mosses springs up. Different species and individual leaves jostle to catch the light as it filters through the plants above.

Climbing plants have their roots in the ground but use taller trees as support. Rattan palms have a barbed stem and use brute force to grow up and hook onto surrounding trees. They can reach a length of 60 metres or more, winding their way up and across different trees, until they are in a good position in the sun.

In Ghana, an area of rainforest covering 0.5 hectares (the size of a large garden) was shown to contain 350 different plant species.

One single tree in the rainforest of West Africa was found to support 47 different species of orchid.

In Costa Rica, the La Selva Forest Reserve contains as many plant species as the whole of Britain, but in an area that is 17,000 times smaller.

Between 50 and 90 percent of the plant species that exist on Earth are found in the rainforests, even though the rainforests cover just over 10 percent of the total land surface area.

Bromeliad

Pike-headed vine snake

Orchids produce some of the most beautiful flowers in the world. Their colour and fragrance attracts insects to come and pollinate them. Only when pollen from another orchid has been placed in the female part of a flower is the orchid able to set seed and reproduce. Some orchids even copy the smell and shape of female insects so that male insects try to mate with the orchid. When they do this they pick up lots of pollen to pass onto the next flower.

Fragrant orchid

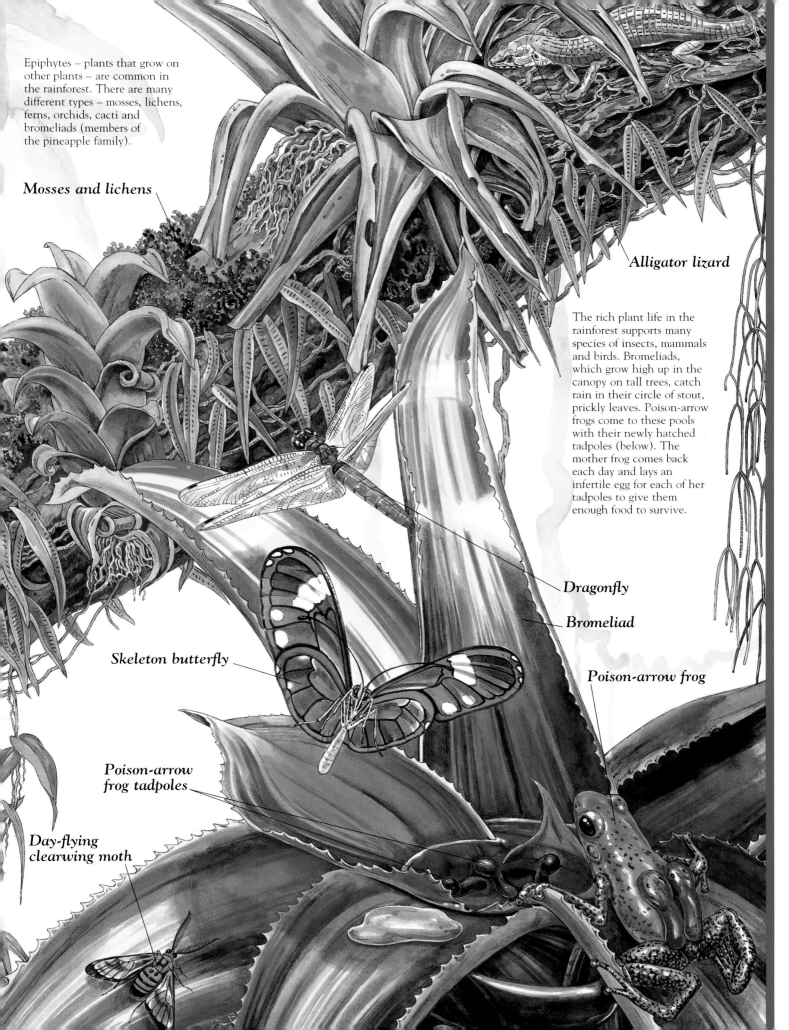

Epiphytes – plants that grow on other plants – are common in the rainforest. There are many different types – mosses, lichens, ferns, orchids, cacti and bromeliads (members of the pineapple family).

Mosses and lichens

Alligator lizard

The rich plant life in the rainforest supports many species of insects, mammals and birds. Bromeliads, which grow high up in the canopy on tall trees, catch rain in their circle of stout, prickly leaves. Poison-arrow frogs come to these pools with their newly hatched tadpoles (below). The mother frog comes back each day and lays an infertile egg for each of her tadpoles to give them enough food to survive.

Dragonfly

Bromeliad

Skeleton butterfly

Poison-arrow frog

Poison-arrow frog tadpoles

Day-flying clearwing moth

The spider monkey uses its strong prehensile tail as a fifth limb to move quickly through the trees and twisted lianas of the rainforest.

Animal Paradise

The world's rainforests are home to 9 out of 10 non-human primates. Forty percent of all birds of prey live in them and a staggering 80 percent of the world's insect population creeps around in the rainforests. As the rainforests are disappearing, many of these precious creatures are becoming endangered. Several species become extinct every day. The primates are one of the most vulnerable groups – some are on the verge of extinction with only a few hundred left alive. The woolly spider monkey, for example, is now extremely rare.

The Emerald tree boa of South America (above right) blends in perfectly with its surroundings. It is a canopy hunter that prowls around at the tops of trees, coiling around branches and lying in wait for the small mammals, birds and frogs that live there. It can grow up to 3 metres long.

The tree porcupine (far right) looks prickly but it is considered a delicacy in West Africa. Hunting such 'bushmeat' is putting great pressure on the survival of some species.

Spider monkey

Squirrel monkey

Emerald tree boa

Iguana

Topaz hummingbird

Passion flower

Heliconid butterfly

Leaf-cutter ant

Hummingbirds drink nectar. Some rainforest plants put so many nutrients into their nectar that it is almost a complete food. Insects that drink the nectar can live on that alone. Larger animals, such as hummingbirds, supplement this diet with a few choice insects that they pick off the flowers they visit.

The Topaz hummingbird (left) sips nectar from passion flowers. The bird picks up pollen on its head as it drinks. When the bird goes on to the next flower, the pollen is brushed onto the female part of the flower and so fertilises it.

The vibrant colours of the red-eyed tree frog (centre) are not just decoration. They warn other animals that the frogs carry potent poisons in their skin. Some native rainforest people use these frogs to load arrows with poison. They scrape the arrow on the frog's skin before going off to hunt.

Tree frogs have very thin skins but they do not dry out high up in the trees because the air in the rainforest is humid and heavy with moisture.

Leaf-cutter ants (left) are an ingenious group of insects. They cut up leaves with their scissor-like mouthparts and then carry them back to their underground nest. The leaves are then chewed thoroughly, but not eaten. Instead the ants 'plant' them with spores from special fungi. The fungi grows, feeds on the leaves and then the ants feast on their crop.

13

Forest People

Yanomami Indians from Brazil and Venezuela cover a large area each day in their search for food and wood for cooking.

The Yanomami diet is high in fresh fruit and plant material but they also eat fresh meat everyday including fish, monkeys and wild pigs.

People have made the rainforest their home for thousands of years. The earliest record of this – from caves deep in the jungle in Borneo – dates back 39 centuries. However, finds like this are rare and most rainforest people have left no trace of their existence. Everything they owned and used, even hardwood arrow heads that could once pierce the tough skin of a wild pig, has rotted away.

Today, 50 million tribal people live in the rainforests. Hunter gatherers like the Mbuti pygmies are nomadic. They travel around constantly, setting up temporary shelters and hunting animals as they spot them. The Siriono of northern Bolivia also wander widely, but for part of the year they clear small plots of land to grow crops. Shifting cultivators like the Iban Sarawak depend more on the crops they grow, using the same piece of land for about 3 years before leaving it. Settled cultivators like the Bantu in central Africa stay permanently in one place and remove the forest cover from the area around them. They are the only type of rainforest dweller to keep herds of domesticated animals.

Shapono

The Yanomami are skilled hunters, using spears, longbows and poison-tipped arrows.

Body painting is very important to the Yanomami. Red is the most common colour but black is also used to indicate bravery or mourning. Other decorations worn include macaw feathers used for head-dresses and armbands. Children pierce their lower lips, noses and ears with sharp sticks.

Members of individual Yanomami tribes live together in a large palm-thatched hut called a shapono (above). At the end of the day, all the Yanomami in the village come together for story-telling, joking and gossiping.

14

Orange-winged amazon parrot

The Yanomami speak many different dialects but they all learn a 'formal' language that all members of each different tribe can understand.

More than 20,000 Yanomami live in the highland rainforests around the border between Brazil and Venezuela. They are the largest group in the Amazon rainforest that still follow a traditional lifestyle. They combine shifting cultivation with hunting and gathering of forest foods. The plantain, a starchy type of banana, is their staple crop.

The Yanomami's territory covers 40,000 square kilometres. Until recently, wars between villages were common, particularly in the centre of the territory. Settlements were highly fortified and groups moved often to escape enemies. Today, the Yanomami's survival is being threatened by road building and mining and they are uniting to demand protection for their land.

From Floor...

The forest floor is a warm, gloomy place. The air is still and everything drips with moisture. In the evergreen rainforests, very little light reaches the ground and only a few shade-tolerant plants are able to grow. The first few metres are dominated by bare trunks and the buttress roots that support them. In more open forests, such as those on the edge of the rainforests that span Indonesia, Thailand and India, there is a thick jungle of young trees, shrubs and lianas.

Because the layer of soil in rainforests is thin, trees can only grow surface roots which barely go underground. These cannot support a tall tree, so extra roots grow from the trunk, arching out to the ground all around the base of the trunk. The aerial roots in mangrove swamps perform the same supporting role, helping to keep the trees from being swept away by the pull of the water.

The forest floor is relatively quiet compared to the canopy and sometimes it is possible to hear frog calls coming from underground. These could be Leptodactylids, a type of frog that has left streams to live in small, damp burrows. Before the rainy season, a male calls to attract a female and, if successful, mates with her in the burrow. The female lays her eggs on his back, into a frothy substance that he has whipped up with water, air and mucus. Inside this frog meringue, the eggs can stay moist until they are ready to hatch and swim out into the forest at the height of the rains.

Glass-wing butterfly

Coati

Most of the flowers and fruits form in the canopy. These attract insects, animals and birds, making this a place that is teeming with life.

...to Ceiling

Thirty metres above the forest floor, intense sunlight shines for many hours everyday onto the top of the canopy. Tall trees laden with epiphytes and climbers branch into a vast green umbrella. Most of the light is trapped by this dense layer of vegetation – less than 2 percent of it ever reaches the ground. The canopy is hot with daytime temperatures averaging 32°C. Although the air is still damp, humidity among the highest branches rarely reaches more than about 60 percent.

The canopy is the powerhouse of the rainforest. The leaves here are the major site of photosynthesis. This is the process by which plants use energy from sunlight to change carbon dioxide and water into simple sugars. The food formed in this way nourishes the plant and also feeds all the animals that eat that plant.

If one of the trees in the canopy dies, light pierces the gloom of the lower levels. Seeds and seedlings react quickly and shoot up to be the first to fill the gap.

The three-toed sloth (centre) is one of the slowest-moving animals on Earth. It not only moves very slowly, appearing to be 'frozen' for hours at a time, but it can also take up to a month to digest its food. Although it would make a good meal for jaguars and other predators, most do not notice the sloth as it hangs motionless and quiet in the trees, high up in the canopy.

Green-winged macaw

Harpy eagle

Glass-wing butterflies

19

By Day

In daylight, the canopy of the rainforest is brightly lit. The light dapples through the leaves and illuminates the magnificent colours of the birds, butterflies, frogs and lizards. Some canopy dwellers use colour to attract a mate, some to blend in with the exotic flowers and fruits. Others use it to warn potential predators that they are inedible or dangerous to eat. There are also the really clever ones that copy warning colours to pretend they carry a large dose of deadly poison.

Many of the primates of the rainforest are in danger of extinction. The woolly spider monkey or miriqui (far right) is the largest primate that lives in the rainforest of South America. Four hundred years ago there were 400,000 woolly spider monkeys – now there are only 11 groups, containing 400 individuals altogether. The government of Brazil and conservation groups are trying hard to save it from extinction.

Some of the most spectacular birds in the world live in the South American rainforests. Keel-billed toucans (above right) have unremarkable bodies topped off with magnificent beaks. Surprisingly, the beak is not as heavy as it looks, since it is made of a light, hard material that is stretched over a web of bony struts. The beak still manages to scare off predators because it looks dangerous and is brightly coloured to highlight its size.

Common iguana

Keel-billed toucan

Red-faced uakari

Saw-billed hermit hummingbird

Red-eyed tree frog

20

By Night

Owl monkey

Hawk moth

Pit viper

Red-eyed tree frog

At night, a totally different cast of characters fills the rainforest stage. Nocturnal animals, like bats and sloths, and night-flying insects wake up. Moths and fruit-bats tend to feed on the most scented flowers, as these are easier to detect in the dark. Many plants have become adapted for this and put their energy into creating scent instead of brightly coloured petals. They rely on visits from nocturnal creatures for their pollination and also for spreading their seeds around the forests.

Olingos (far left) are mammals from the same family as the raccoon and the panda family. They are slender, 33–46 cm long, with a very furry tail 38–51 cm long. Olingos live in the rainforests of Central America and eat mainly fruit.

In order to reach its food, the silky anteater (bottom left) scratches a hole in an anthill with one of its sharp, curved claws. It darts its tongue in and out of the hole capturing the ants and swallowing them whole. Though anteaters never attack another animal first, they will defend themselves fiercely. When in danger, they will fight off attackers with their thick, strong claws which can be as long as 10 cm, making anteaters a match for even mountain lions or jaguars.

Red-eyed tree frogs (left) come from the rainforests of Costa Rica in South America. They have red eyes as a method of defence against predators. If a frog is disturbed by a predator while asleep, the frog's eyes pop open. This sudden display of brightness startles the predator, as such large staring eyes could be those of an enemy, poised to attack. A moment's hesitation from the attacker is all the agile tree frog needs to leap to safety.

Undiscovered Treasures

Bromeliad

Tayra

The rainforest is not just an area of great beauty. It is also a cradle of life, a home to hundreds of people and a place where many rare species of plants and animals exist. Without this 'store' of life, we do not know what might happen in the future. One area that is already losing out is medicine. Only a few plants have been tested but many have already been found to contain substances to treat cancer or AIDS. How many more potential treatments have we lost already by clearing the forests?

The hoatzin (far right) is one of the few leaf-eating forest birds of South America. The leaves can stay in the bird's stomach for two days while being digested by bacteria. This might explain why the bird smells like cow manure and flies very badly.

Death and Destruction

The world's ever increasing human population puts great pressure on the rainforests. By the year 2100, eight billion people will be living in countries that contain rainforests. In order to feed their people, these countries will need more farming land. That means more areas of rainforest will be cut down, cleared or burned. More trees will be destroyed as fuel. The rivers running through the forests will be dammed to generate electricity and to allow roads to be built. Underground mines may be constructed to obtain the rich natural resources below. If such exploitation of the rainforests goes unchecked, they may never recover.

The rainforests are part of the Earth's life support systems. They affect weather patterns and the trees hold the nutrient-rich soil together, preventing erosion. When hectares of forest are cut down, flooding increases and large areas of soil are washed away. Once the soil is gone, no amount of human effort can re-establish the unique environment of plants and animals that made up the rainforest.

Hope for the Future

Rainforest that has been plundered can recover if it is left alone. New plants grow quickly to reclaim cleared land and the thick new vegetation soon becomes filled with insects and small mammals. But leaving the rainforest alone is not really an option. Our need for its resources is too great. We have to think about how to use the forests in a sustainable way. That means taking what we want without destroying what it has to offer for future generations. We must remember that the rainforests are irreplaceable and invaluable to life on Earth.

We still need many of the products available from the rainforests – timber, latex, fruit, rattans, medicinal plants, rubber, oil, fibres and cellulose. Latex is tapped from wild rubber trees by cutting slanting grooves in the trunk and letting the sap ooze out (above right). Plantations can provide these materials, without causing soil damage, if they are managed carefully.

Some parts of the rainforests must be kept untouched. Nature reserves have already been set up in rainforests all over the world (see right). There is one reserve for each type of rainforest ecosystem. These areas are protected but can still be studied to find out more about the value of the rainforests.

Stink bug

28

Oil wells (left) are an ugly sight in the middle of a rainforest but they do surprisingly little damage. Only small areas have to be cleared for them and then most of their activities go on below ground. Oil is also a very good source of income. Some of the money it makes can be used to fund conservation projects.

Many wildlife conservation projects are succeeding. In 1972, India set up 14 tiger reserves as part of Project Tiger. At that stage, fewer than 2,000 tigers were left in the country. Today, 24,700 square kilometres has been put aside as a protected habitat for the tigers. Their number has stopped falling and is just starting to rise again.

The right of native people to live as they have lived for generations is now being recognised. In some projects, field workers from large companies are working with the shaman (medicine man or woman) in different tribes to identify rainforest plants that are used in traditional medicines. If these are developed as modern medicines, some of the profit made is given back to the tribe who use it to protect their tribal lands.

A shaman of the Kampa tribe (below left) is preparing ayahuasca, a hallucinogenic substance made from the bark and leaves of certain trees. In the West, its ingredients have been used to treat malaria.

White-tailed deer

Katydid

Glossary

Aerial root
Roots that stick out from the trunk of a tree, above ground level. Aerial roots have two functions: to prevent the roots becoming rotten by being in water all the time; and as support for a tall tree.

Altitude
The height of ground above sea level. The top of mountains are at high altitude. Coastal areas are at low altitude.

Canopy
The uppermost layer of the rainforest.

Conservation
The protection and maintainance of an environment, sometimes aimed at one particular species of animal or plant.

Deforestation
The clearing of areas of trees or forests.

Ecosystem
A living system that includes animals and plants.

Epiphytes
Plants that use other plants for support. Their roots never reach the ground.

Equator
The imaginary line around the centre of the Earth that indicates the parts of the planet that are always closest to the Sun.

Evergreen
Evergreen trees and plants are those that never lose their leaves. They remain green and in full growth, all year round.

Extinct
A plant, animal or any other living thing that has died out.

Habitat
The natural home of a living thing.

Hallucinogenic substance
Something that produces vivid waking dreams and a trance-like state.

Humidity
A measure of how much moisture there is in the air. High humidity makes people feel hot, clammy and sweaty.

Liana
A type of climbing plant that grows in tropical forests.

Nocturnal
A creature that is active during the night but sleeps during the day.

Nomadic
Nomadic people move constantly from one region to another, usually taking their domestic animals with them. They rarely settle in one place for long.

Photosynthesis
A chemical reaction that plants use to feed themselves. They combine carbon dioxide from the air with water from the soil to make energy-rich sugars such as glucose. The energy to power the process comes from sunlight, and plants release oxygen as a by-product.

Poison
A substance that can cause injury or even death if eaten or taken in by a living thing.

Pollination
The sprinkling of pollen onto a particular part (the stigma) of a flower.

Predator
A type of animal that preys on other animals for food. Jaguars and emerald tree boas are both predators.

Prehensile
This means 'like a limb'. Many monkeys in the rainforest have a prehensile tail that winds round branches, supports their weight and behaves like another limb.

Primates
The name for a sub-division of the animal kingdom that includes animals such as apes, monkeys and people.

Rattan
A species of climbing palm.

Settled cultivator
People who settle permanently and set up farms.

Shifting cultivator
People who stay in one place for a while to grow crops but who then move on to a new site after two or three years.

Soil erosion
The wearing away of the soil. Soil erosion is common in rainforest areas where the trees have been cut down. With no roots and vegetation to hold the soil together, it is easily blown or washed away by wind and rain. In this way the area loses its nutrient-rich soil and the land becomes dead.

Sustainable
Using a resource in a sustainable way means using it without using it all up. Planting fast-growing trees to replace those cut down for timber is a sustainable way of using forests. Cutting down old, slow-growing wild trees is not.

Swamp
An area of waterlogged ground.

Vegetation
All types of plants and plant life.

Rainforest Facts

Most tropical rainforests have an annual rainfall of at least 250 centimetres and are filled with tall broad-leaved evergreen trees that form a continuous canopy.

Rainforest products, including rattan, bamboo, woods such as rosewood and mahogany, nuts and spices are sold around the world.

Deforestation (the destruction of trees) accounts for the loss of nearly two-thirds of Central America's rainforests – over 400.000 hectares every year.

The most species-rich plot of rainforest known is in Peru: 283 species of trees were found in just one hectare. Every second tree is a different species. While this particular area is the most species-rich, this kind of growth and diversity is typical of most rainforests.

In Sarawak, Borneo, the Penan people use over 50 medicinal plants which they harvest from the forest. The plants are used for poison antidotes, contraceptives, clotting agents, general tonics, stimulants, disinfectants, remedies for headaches, fever, cuts and bruises, boils, snake bites, toothache, diarrhoea, skin infections and rashes, and for setting bones.

Popular woods such as mahogany or rosewood, used for the manufacture of furniture, are only found in the rainforests.

Most of the nutrients in a rainforest ecosystem are stored in its vegetation rather than in its soil.

One of the biggest industries throughout Central America is cattle ranching. Most of the beef produced is exported to North America for use in fast-food restaurants. The most common way to clear land for ranches is to tear down and set fire to the trees, a practice known as 'slash-and-burn' agriculture. On 9 September 1987, a satellite picture of the Amazon River Basin showed a total of 7,603 fires burning in the rainforest.

The rate of rainforest destruction is such that species of plants and animals are becoming extinct and disappearing from the rainforests before they can be studied.

About 2,000 trees per minute are cut down in the rainforests around the world. In most of the countries with tropical rainforests, only one tree is replanted for every 10 cut down. In some countries the rate is one tree replanted for every 30 destroyed.

The largest threat to the survival of jaguars is the loss of the rainforests they live in. Their numbers have fallen fast and they have already disappeared completely from many of their former regions. The only places they are still relatively common are remote areas of Guatemala and Belize in Central America.

Since the turn of the century, 90 tribes of native people have been wiped out in the Brazilian forests. Twenty-six of those tribes were killed or dispersed in the past decade alone.

In the Himalayas, in Nepal, about a quarter of a million tonnes of topsoil are eroded away every year. This is due directly to the removal of the forests from that region. Even more soil is lost from the Himalayan foothills in India.

Western medical uses for plants discovered in rainforests include: a medicine for malaria (the bark of the cinchona tree produces quinine); a muscle relaxant used during surgery (curare, a vine extract used by native peoples to poison arrows and darts); and a treatment for depression (secretions of a particular Amazonian frog).

Due to its specialised soil, rainforests are unsuitable for growing any crops such as wheat or vegetables. Only trees and specialised vegetation thrive in such areas.

No one knows just how much the rest of the global ecosystem depends on rainforests, but we may find out in the next 30 to 50 years. That is how long it is estimated that it will take for tropical forests to disappear altogether, if current trends continue.

In the rainforests, there are some truly extraordinary plants. There is a fruit with more vitamin C than an orange, a palm with more vitamin A than spinach and another palm whose seeds contain 27% protein. One type of tree produces a resin that can be used unprocessed to run a diesel engine, while another yields up to 300 kilograms of oil-rich seeds a year.

31

Index

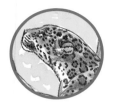